T0075454

Lecture Notes in Mobility

Series Editor

Gereon Meyer, VDI/VDE Innovation und Technik GmbH, Berlin, Germany

More information about this series at http://www.springer.com/series/11573

Jan Brinkmann

Active Balancing of Bike Sharing Systems

Jan Brinkmann
Institut für Wirtschaftsinformatik
Technische Universität Braunschweig
Braunschweig, Germany

ISSN 2196-5544 ISSN 2196-5552 (electronic)
Lecture Notes in Mobility
ISBN 978-3-030-35011-6 ISBN 978-3-030-35012-3 (eBook)
https://doi.org/10.1007/978-3-030-35012-3

This Springer imprint is published by the registered company Springer Nature Switzerland AG
The registered company address is: Gewerbestrasse 11, 6330 Cham, Switzerland

Foreword

Vehicle sharing has received a remarkable attention as a new means of urban transportation. Practice has shown that the one-way use of vehicles follows mobility patterns of people leading to temporal and spatial imbalances with respect to the distribution of vehicles in the city. In station-based bike sharing systems, customers suffer from the absence of bikes in case of a potential rental and the absence of bike racks in the case of a bike return. Station-less systems have claimed to offer flexibility; however, they have failed to overcome the deficiency of bike imbalances.

System operators see the requirement of redistributing bikes between city areas over the day at significant expenses. A methodological support of bike logistics has concentrated on static optimization models. These models are typically fed with data of historic bike usage. Since history does not repeat itself, optimal solutions obtained from static model cannot be implemented due to stochastics with respect to actual bike usage.

Jan Brinkmann focuses on a control approach deciding dynamically about bike imbalances to be resolved. He combines control with an anticipation of future redistribution demand by means of online simulation. The simulation takes into account the driving time needed to arrive at the respective station, the loading or unloading time at this station as well as the avoidance of future fails resulting from bike inventory changes.

The informative value of the simulation strongly depends on the simulation horizon. A short horizon may not reflect the utility of the station visit. Simulating over a long horizon may report on customer fails, which no longer relate to the respective station visit. Jan Brinkmann is able to provide evidence that a suitable simulation horizon is by no means fixed, but depends on the particular situation, i.e., the time of day. To this end, he develops an approximate dynamic programming approach determining heterogeneous simulation horizons iteratively.

The above consideration applies to the one vehicle case only. Whenever a fleet of trucks is employed for bike redistribution, the decentral decisions of the trucks are no longer independent of each other. Since all of them follow the same decision model, it may happen that demanding stations may accidentally be visited multiple times. Jan Brinkmann suggests different levels of coordination coming along with a slightly growing need for information exchange. The trucks operate independently of each other and take decision for their own operation. Like in the one vehicle case before, decisions comprise the number of bikes to be loaded or unloaded at the current station and the station to be visited next.

The control approaches developed are carefully validated for real-world instances of bike sharing systems. Promising results are obtained for all instances considered. In particular, the approach is best suited for bike sharing systems which do not show a regular structure of bike imbalances due to commuter travel. Regular flows from residential areas to office districts in the morning and reverse flows in the late afternoon are relatively easy to predict and to counteract. More challenging are complex mobility patterns consisting of mixed work, shopping, and leisure activities. Results obtained indicate that these complex interactions can be supported much better by control than by static optimization.

Jan Brinkmann pioneers online control models for the redistribution logistics of bike sharing systems. The work bases on a solid understanding of bike sharing system, business models, and related activities. The control approach pursued has been well received by the transportation research community as well as by colleagues working in Operations Research. This book summarizes research of recent years by giving a comprehensive introduction into control approaches for today's and forthcoming vehicle sharing systems.

Braunschweig, Germany
January 2019

Dirk C. Mattfeld

Preface

Many cities suffer from discomforts caused by individual and motorized traffic. Therefore, city administrations implement sustainable shared mobility services such as bike sharing systems (BSSs). In BSSs, users are allowed to rent and return bikes on short notice at stations. Data analysis reveals that rental and return requests follow spatio-temporal patterns such as commuter usage and leisure activities. In the morning, commuter usage is indicated by mainly rental requests in residential areas and mainly return requests in working areas. This behavior inverts in the course of the day. The resulting unequal requests lead stations to become empty or full. Requests to rent bikes will fail at empty stations. At full stations, requests to return bikes will fail.

Providers counteract these inconveniences by means of balancing. In this work, we focus on the operational management's view on the balancing of BSSs. That is, the provider schedules transport vehicles relocating bikes between stations in order to minimize the amount of failed requests. As requests are uncertain, the resulting challenge is to identify stations with a lack or a surplus of bikes. To this end, we introduce approaches simulating future requests and approximating expected amounts of failed requests. Then, anticipation is enabled by means of including the approximations in the decision making process.

We evaluate our approaches in case studies based on real-world data. The results point out that our approaches are able to reduce the amount of failed requests significantly compared to common benchmarks from literature.

Braunschweig, Germany
January 2019

Jan Brinkmann

Contents

Acronyms

ADP	Approximate Dynamic Programming
AV	Autonomous Vehicle
BSS	Bike Sharing System
DLA	Dynamic Lookahead Policy
DPS	Dynamic Policy Selection
IRP	Inventory Routing Problem
LA	Lookahead Policy
LUT	Lookup Table
MDP	Markov Decision Process
PTS	Public Transport System
SMS	Shared Mobility System
SLA	Static Lookahead Policy
STR	Safety Buffer-tending Relocation Policy
VFA	Value Function Approximation
VRP	Vehicle Routing Problem

List of Figures

List of Tables

List of Algorithms

Chapter 1
Introduction

Increasing urbanization and mobility demand lead to a large volume of traffic in urban areas. As the traffic infrastructure is limited, too much traffic results in traffic jams, noise, and air pollution. City administrations focus on a reduction of the individual traffic to tackle these discomforts. Therefore, collective traffic modes, i.e., public transport systems (PTSs), are launched or expanded. Conventional modes are buses and trams. On the one hand side, PTSs may be able to reduce urban traffic. But on the other hand side, the comfort is reduced. First, the users' actual origins and destinations are not necessarily in walking distance to a bus or tram station. Second, buses and trams are often delayed or crowded.

Shared mobility services (SMSs), such as car, bike, and scooter sharing systems, are promising alternatives as well as complements to conventional PTS. An SMS grants the access to an available wheeler where the user does not become the owner. The access is granted for one trip, i.e., for the time span the user needs to drive from his origin to his destination. A trip can be started any time a car, bike, or scooter is available without restrictions due to timetables. When the user ends his trip, it becomes available to other users. In this way, car sharing systems can reduce the number of cars in the city significantly (Archer 2017). Further, bike sharing systems (BSSs) offer emission-free and sustainable transport. In free-floating BSSs, bikes are distributed in the operation area. In station-based BSSs, the access to bikes is granted at stations, i.e., bikes are rented and returned at predefined locations. Stations are capacitated, i.e., a limited number of bike racks is available. The network of stations expands over the city center as well as residential areas. Thus, users can satisfy there mobility demand completely with the BSS if origin and destination are in cycle distance. BSSs serve as a complement to PTS if the next PTS stations is not in walking distance. Then, a user can rent a bike near his origin and can cycle to the next PTS station (Lin and Yang 2011).

Information technology found its way into BSSs to support user authentications and payments, and to record rental and return requests (Vogel 2016). The resulting availability of data allows data analysis to reveal spatio-temporal patterns of requests. BSSs are often used by commuters. Accordingly, in the morning, commuters request to rent bikes in residential areas and request to return bikes in working areas. In

J. Brinkmann, *Active Balancing of Bike Sharing Systems*, Lecture Notes in Mobility,
https://doi.org/10.1007/978-3-030-35012-3_1

the evening, bikes are rented mainly in working areas and returned in residential areas. Additional requests are due to leisure activities (Froehlich et al. 2009; Vogel et al. 2011; O'Brien et al. 2014). Unfortunately, unequal amounts of rental and return requests cause inconvenience. In free-floating BSSs, some areas may run out of bikes and users are detained from renting bikes. In station-based BSSs, stations become empty or full, respectively. No bikes can be rented at empty stations, and no bikes can be returned at full stations. Therefore, the provider's supply does not match the users' demand. In other words, the BSS is imbalanced.

The scope of this work is active balancing of bike sharing systems. In general, balancing aims on consolidating supply and demand (Sieg 2008). For BSSs, that is, to offer a bike when a user requests to rent one and to offer a free bike rack when a user requests to return a bike. Requests are uncertain to the provider but follow a recurring pattern. The balancing of BSSs is addressed by various management layers differing in the planning horizon. Long-term decisions concerning the network of stations, i.e., locations and capacities, are made by the strategical management. Both, the tactical and operational management aim on scheduling transport vehicles to manually relocate bikes between stations and, in this way, to *save* requests. Vehicles pick up bikes at stations with a lack of free bike racks. Bikes are delivered to stations with a lack of bikes. Here, the tactical management schedules master tours that are realized every day (mid-term). Master tours are efficient if the majority of requests can be forecasted. However, rescheduling of vehicles becomes necessary if a significant amount of requests occurs unexpected. The operational management aims on active balancing of the BSS. That is, first, to anticipate requests, and second, to counteract unexpected requests by means of appropriate (re)scheduling. In this work, we mainly discuss the operational balancing of BSSs to support active relocationing.

We depict the operational balancing as a stochastic-dynamic multi-vehicle inventory routing problem for BSSs ($\mathrm{IRP}_{\mathrm{BSS}}$). An $\mathrm{IRP}_{\mathrm{BSS}}$ is a vehicle routing problem (VRP) where decisions concerning sequences of stations as well as stations' inventories are made. It is stochastic since request are uncertain but follow a stochastic distribution. It is dynamic since multiple decision points are considered over the day to counteract unexpected requests. Therefore, we model the $\mathrm{IRP}_{\mathrm{BSS}}$ as a Markov decision process (MDP). In the MDP, the stations are characterized by their fill levels, i.e., the number of bikes currently located at the station, and a maximum capacity, i.e., a number of bike racks. In the course of the day, successful requests alter the fill levels. If a station's fill level is equal to zero, rental requests will fail. If the station's fill level is equal to its capacity, return requests will fail. If a request fails, the associated user approaches a neighboring station. When a vehicle arrives at a station, we have to determine the amount of bikes to be picked up or to be delivered at these station as well as where to send the vehicle next. Then, the decisions are realized. Policies return decisions with respect to the current state of the BSS. The objective is to identify an optimal policy minimizing the expected amount of failed requests.

Solving the $\mathrm{IRP}_{\mathrm{BSS}}$ to optimality requires non-polynomial time. Therefore, we draw on methods of approximate dynamic programming (ADP) to enabling applicability for real-world instances. More specific, we use lookahead policies (LAs) and data of recorded requests. LAs anticipate future requests. That is, to incorporate future requests in the decision making process. Therefore, an LA evaluates potential decisions by means of simulations over a lookahead horizon before selecting the decision that saves the most requests. To this end, the progresses of stations' fill levels (realized by potential decisions) are simulated. In this way, resulting failed requests are approximated. We propose two approximation methods: An offline lookahead policy (LA_{off}) successively updates the stations' fill levels with average magnitudes. This method is computational cheap but neglects interactions between stations. By contrast, an online lookahead policy (LA_{on}) draws on requests resampled from a stochastic function representing the spatio-temporal distribution of requests. The MDP itself is simulated for every set of resampled requests. In this way, the progresses of fill levels and, therefore, failed requests can be observed while interactions between stations are taken into account. Although, an LA_{on} requires a multiple of the computational power, it is still applicable for real-world instances.

When taking the decision for the vehicle under consideration, we maximize the amount of saved requests by anticipating future decisions of the whole vehicle fleet. That is, we recall the approximated failed requests and determine the amounts of requests the individual vehicles can save at the different stations. Here, we take the locations and the amounts of bikes currently loaded by the vehicles into account. In the resulting assignment problem, every vehicle is assigned to one station. The goal of the assignment problem is to maximize the amounts of saved requests over all vehicles. However, we realize the assignment of the current vehicle only. By doing this, we preserve flexibility to counteract unexpected requests.

Regarding the length of the LAs' lookahead horizon, it turned out that, the horizon needs a minimum length in order to include a sufficient number of requests in the approximation process. However, a limitation of the horizon is beneficial since the approximations may become inaccurate and the vehicles may be misguided. Due to the spatio-temporal pattern of requests, the suitable length of the horizon depends on the time of the day. Therefore, we draw on dynamic lookahead policies (DLAs). For the sake of simplicity, we partition the day into periods. Then, we define a sequence of horizons, i.e., one horizon for every period. To this end, we apply a value function approximation (VFA). The VFA approximates the impacts of different horizon lengths in certain periods. In this way, the VFA adapts a sequence of horizons to the distribution of requests.

In computational studies on real-world data by the BSSs of Minneapolis (MN, USA) and San Francisco (CA, USA), we compare the LAs with myopic as well as anticipatory benchmark policies from literature. The results point out that applying DLAs is highly beneficial compared to LAs with static horizons and the benchmarks.

Figure 1.1 provides an overview on the parts and chapters of this work.

The parts read as follows:

Part I: Preliminaries

The first part covers the preliminaries and comprises three chapters. In Chap. 2, we discuss urban mobility, benefits, functionalities, typical requests patterns, and the three management layers of BSSs. A review on related optimization problems is

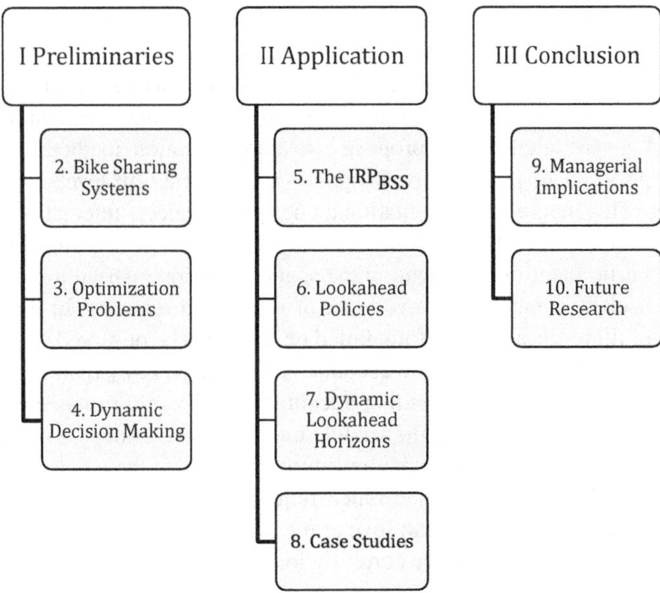

Fig. 1.1 Overview on the parts and chapter of this work

given in Chap. 3. We distinguish VRPs for various applications and IRPs for BSS. The basics on dynamic decision making are defined in Chap. 4. Here, we introduce the modeling framework MDP and the class of solutions methods ADP.

Part II: Application

Four chapters are allocated to the second part on the application of the dynamic decision making framework for the IRP_{BSS}. In Chap. 5, the notation of the IRP_{BSS} and the corresponding MDP are defined. The LAs for solving the IRP_{BSS} are introduced in Chap. 6. In Chap. 7, we define the VFA to determine dynamic lookahead horizons for the LAs. Chapter 8 comprises a case study based on real-world data showing evidence of the MDP's and LAs' applicability.

Part III: Conclusion

In the third part, Chap. 9 highlights the managerial implications derived from our findings. Future research is discussed in Chap. 10. Here, we point out various issues of urban logistics that are strongly related to the balancing of BSSs but are not considered in the field of BSSs yet. Additionally, we discuss different methods of ADP that may enhance computational decision making.

Part I
Preliminaries

Chapter 2
Bike Sharing Systems

In this chapter, we introduce the concept of BSSs and point out the resulting challenges. In Sect. 2.1, we discuss urban mobility and show how BSSs are integrated. In Sect. 2.2, we describe the benefits BSSs offer. The functionalities of BSSs are presented in Sect. 2.3. Typical request patterns and resulting challenges are discussed in Sect. 2.4. Section 2.5 addresses the management of BSSs.

2.1 Urban Mobility

An increasing population worldwide (United Nations 2017) and an increasing urbanization rate (Dong et al. 2017) result in growing cities. An also increasing mobility demand results in high volumes of individual motorized traffic in urban areas. Therefore, urban areas suffer more and more from environmental pollution such as carbon dioxide, nitric oxide, and noise as well as from traffic jams. To counteract these inconveniences, local administrations implement PTSs. A PTS is a public mode of human transportation for collective trips (Rodigue et al. 2013). PTSs are realized with different modes, e.g., buses, streetcars, or subways. These modes drive along a route with stations. At the stations, users can enter or leave the mode. Typically, the network of stations is close-mashed in the city center and coarse-meshed in the outskirts which leads to discomforts if a user's actual origin and/or destination are not in walking distance to a station. This circumstance is know as the first and last mile problem (Shaheen et al. 2010). Additionally, the modes follow timetables. That means that stations are visited in certain time intervals. Consequentially, users cannot start their trips where and when they want. Further, punctuality is rarely observed.

The first and last mile problem can be counteracted by means of BSSs. According to Meddin (2017), 1,500 different cities around the globe run BSSs. BSSs are public rental services providing bikes to users. In BSSs, users can use bikes on short notice in an unspecified time span and for self-determined mobility. A trip on a bike starts

J. Brinkmann, *Active Balancing of Bike Sharing Systems*, Lecture Notes in Mobility,
https://doi.org/10.1007/978-3-030-35012-3_2

with the *rental request* at some place in the city. Then, the user is free to use the bike. The trip ends with the *return request*—most likely—not at the place where the bike has been rented. Thus, a user can rent a bike near his home and can return it at his working place, at a PTS station or at the main railroad station (Büttner et al. 2011). The permission to rent a bike is usually granted by the membership in the BSS. The membership comes with an annual, monthly, or weekly fee. Members are allowed to rent bikes as often as they want. Typically, trips of less than 30–60 min are free of an additional charge. For additional time, providers calculate a price of a certain graduation. Further, non-members are granted access if they possess a credit card (Midgley 2009). For more details on the business model of BSSs, we refer to Vogel (2016).

2.2 Benefits

BSSs offer a number of potential benefits for the associated cities and especially for the city's inhabitants. We identified three categories of benefits: reduction of traffic (Sect. 2.2.1), improvement of health (Sect. 2.2.2), and increase of tourist attraction (Sect. 2.2.3). In the respective sections, studies are introduced to provide evidence of the benefits.

2.2.1 Reduction of Traffic

Fishman et al. (2014) analyze the substitution rate of cars against bikes induced by the local BSSs in different cities. The authors undertook investigations for the cities of Brisbane (Australia), Melbourne (Australia), London (UK), Washington D.C. (USA), and Minneapolis (MN, USA). In a survey, the members of the BSSs have been asked about their estimated behavior in the absence of the BSS. More precisely, the members should think about their last trips with the BSS's bike and have been asked about the mode they would have used if the BSS would not have been available. In this way, the authors estimate the ratios of car/bike substitution and reduction of traffic. The survey revealed heterogeneous results. 21% of the members of Brisbane, and 19% of the members of Melbourne and Minneapolis would have used a car instead of a BSS's bike. In London and Washington D.C, only 2 and 7% of the members would have used a car instead. The authors state that the different ratios are due to the commuter car usage. In Brisbane, Melbourne, and Minneapolis, 70–76% of the people working in the associated cities go to work by car. For London and Washington D.C., the rates are in the range of 36–46%. Therefore, BSSs decrease the usage of private cars.

Wang and Zhou (2017) investigate the impact of BSSs on traffic congestions. In a study on traffic data by 96 US American cities, the authors identify a dependency between the size of the population in the associated city and the congestion in urban areas and in rush hours. Generally speaking, a larger population comes with more congestion. In the presence of a BSS, the dependency is significantly smaller. Thus, the authors conclude that BSSs decrease the congestion in rush hours.

Shaheen et al. (2011) undertake a survey among the users of the BSS in Hangzhou (China) regarding how the BSS changed their behavior. 41% of the survey's participants responded that their use of the local PTS increased due to the BSS. Further, 62% of the participants use their private car less often and 37% postponed buying a car. The numbers point out that the BSS reduces urban traffic.

2.2.2 *Improvement of Health*

The high CO_2 concentration in the atmosphere is the major cause for global warming (Houghton et al. 1990). Therefore, Massink et al. (2011) estimate the potential CO_2 emission reduction if fossil-fuelled modes are replaced by bikes. The authors draw on a survey conducted by the local authorities of Bogotá (Columbia). In the survey, the participants are asked about their mobility habits including the purpose, the used mode, and the trip length. In a first step, the authors estimate which trips on motorized modes can be substituted by bikes. On this base, a potential reduction of 55,115 tons CO_2 emission per year is estimated.

Rabl and de Nazelle (2012) investigate the impacts on the mortality when individuals change from car to bikes. Benefits are the individual health gain due to physical activity and the collective health gain due to lower air pollution. However, drawbacks are the individual pollution burden, e.g., when cycling behind a bus, and the risk of accidents. The distinct impacts are estimated for datasets from Paris (France) and Amsterdam (The Netherlands), and compared by monetary values. It turns out that the individual health gain by physical activity represents by far the largest impact. The second largest impact is the risk of accidents. Notably, the risk in Amsterdam is much smaller than in Paris due to the cycling infrastructure.

2.2.3 *Increase in Tourists Attractiveness*

So called bike tourism in rural areas has been investigated by Pratte (2006). In a case study on the "Minnesota's State Trails" in the USA, the author shows the requirements of enabling bike tourism. To the best of our knowledge, the tourists attractiveness due to bike usage in urban areas has not been directly analyzed yet. However, Vogel et al. (2011) reveal tourist usage by means of data analysis. In the BSS of Vienna (Austria), stations near tourist hotspots are characterized by a demand pattern which is typical for tourists. O'Brien et al. (2014) state that the BSSs in Rio de Janeiro (Brazil) and Miami Beach (Florida, USA) are even dominated by tourists. This portends to an increase in tourist attraction due to BSSs. For more information on tourist usage, we refer to Sect. 2.4.3.

We conclude, that BSSs lead to manifold and significant benefits. Therefore, we go into details of the functionality in Sect. 2.3.

2.3 Functionality

We distinguish two approaches to realize BSSs. These approaches differ in the way of how users gain access to the BSS's bikes. In free-floating BSSs (Sect. 2.3.1), bikes are available in a defined operating area. In station-based BSSs (Sect. 2.3.2), bikes are rented and returned at stations.

2.3.1 Free-Floating

The idea of free-floating BSSs is that users can rent and return bikes anywhere within an operating area. The bikes can be located, e.g., by means of the global positioning system (Hofmann-Wellenhof et al. 2012). Further, users are able to locate available and near bikes via smartphone apps. When a user finds a bike he likes to rent, the app provides a transaction code to unlock the bike and to start the trip. In some free-floating BSSs, bikes can be reserved for a short time-span of around 15 min. When the user reaches his destination, the bikes can be parked anywhere in the operating area. The bike is returned when it is locked by the user. Then, the bike becomes available to other users (Pal and Zhang 2017).

2.3.2 Station-Based

In station-based BSSs, stations are distributed in an urban area. Every station is equipped with a self-service terminal and has a limited number of racks where bikes can be parked. When a user wants to rent a bike, he can find a station via smartphone app. However, the bike stations usually are eye-catching due to showy colors. The smartphone apps also provide information on the fill levels of stations, i.e., the number of bikes currently available at every station. At the stations, users identify themselves with a membership or credit card. Then, one of the available bikes is unlocked and the user can start the trip. The trips end with the return of the bike at the same or some other station. To do so, the bike needs to be placed in an available rack (Büttner et al. 2011). Figure 2.1 depicts an exemplary station of the BSS in Stuttgart (Germany) where two users handle the terminal in order to rent bikes.

Both free-floating and station-based BSSs have advantages and drawbacks. From the providers' perspective and regarding the implementation cost, free-floating BSSs are advantageous. Here, the existing infrastructure of the associated city is used. Therefore, only the bikes need to be purchased. In station-based BSSs, the stations need to be embedded in the existing infrastructure. This leads to monetary costs and, further, blocks valuable space in the city. Also from the users' perspective, free-floating BSSs seem to be advantageous as bikes can be rented and returned at almost any place. However, in the next section, we will see that additional challenges arise due to the users' request patterns.

Fig. 2.1 A station of a bike sharing system

2.4 Request Patterns

Information technology grants insights into the requests patterns of BSSs. In free-floating BSSs, trip data is generated by tracking the movements of bikes. In station-based BSSs, requests at stations are recorded. As anonymized trip data is often offered to public or is made available for academic usage, BSSs attract research on both data analysis and optimization. Here, we refer to a number of studies analyzing data and drawing conclusions on requests patterns. The studies draw on different BSSs. Nevertheless, a significant amount of conclusions can be generalized.

Froehlich et al. (2009) draw on data recorded in 2008 by the BSS in Barcelona (Spain). Vogel et al. (2011) analyze data by the BSS in Vienna (Austria) from the years 2007 and 2008. O'Brien et al. (2014) conduct a study, based on 38 BSSs in the Americas, Asia, Australia, Europe, and in the Middle East. Reiss and Bogenberger (2015) carry out a study on the free-floating BSS in Munich (Germany).

We do not want to conceal that there are stations where the amount of rental and return requests are more of less even and, therefore, imbalances are rare (Vogel et al. 2011). Other patterns are easy to understand. For instance, Froehlich et al. (2009) observe hardly no returns at stations on hills. Apart of these, we introduce a number of consistent and unique patterns revealed by the referenced studies.

2.4.1 Seasons and Weather

The usage of all BSSs is strongly subject to seasonal variations. This is indicated by the amount of trips in the course of the year. The amount of trips in the summer months is significantly higher than in the winter months due to weather conditions such as the temperature and precipitation. Therefore, the authors limit their data sets to the summer months of the associated BSS. Notably, there is no consistent definition of summer and winter months. Seasons vary with the latitudes of the associated locations.

2.4.2 Commuters

Commuter usage is denoted by a request peak in the morning, when users leave their homes and cycle to work, and by a request peak in the evening, when users cycle back home. The authors of all studies agree that commuter usage is common for most BSSs.

To prove commuter usage, Reiss and Bogenberger (2015) subdivide the operational area of the free-floating BSS into zones and determine the amount of rentals and returns for every zone and for different time periods. Then, heat maps are created where every zone is colored according to the amount of either rentals or returns. The authors compare the heat maps and conclude that commuter usage is significant. Additionally, they state that in summer, commuter usage is less affected by the weather than other purposes.

Froehlich et al. (2009) and Vogel et al. (2011) reveal commuter usage by means of cluster analysis including both spatio and temporal aspects. They analyze the amount rental and return request in the course of the day for every station. It turns out that at stations of a certain cluster, mostly bikes are rented in the morning and returned in the evening. These stations are called residential stations. On the other hand side, at stations of another cluster, mostly bikes are returned in the morning and rented in the evening. The stations in this cluster are called working stations. The findings are undergird as the associated stations are found in residential or working areas, respectively.

According to expectations, commuter usage is observed on working days. Notably, some exceptional BSSs are located in Asia where commuter usage also takes place on weekends (O'Brien et al. 2014).

2.4.3 Leisure and Tourists

The cluster analysis conducted by Vogel et al. (2011) further reveals leisure and tourist usage. They identified stations that dominate all other stations of the associated

BSS in terms of amount of requests between noon and afternoon. The stations are located directly near tourist hotspots. According to Froehlich et al. (2009), request peaks around noon are due to general leisure activities, e.g., when people leave their workplaces to have lunch.

In a summary, it can be said that requests are subject to spatio-temporal variation. Results of the introduced patterns are unequal amounts of rental and return requests and, further, unequal availabilities of bikes and free bike racks. Some stations tend to run out of bikes or while other reach their capacities. At empty stations, rental requests fail as there are no bikes available. At full stations, no return requests can be satisfied as no free bike racks are available. Gauthier et al. (2013) name success factors that every BSS need to fulfil in order to maximize the users' satisfactions. The reliable availability of both bikes and free bike racks is the most crucial factor. User are eager to use a BSS if they can rely on the reliable availability of bikes and free bike racks.

2.5 Management Layers

From the traditional economics' view, demand and supply meet at the market. The market is balanced if the demand equals the supply (Sieg 2008). Following this idea, user requests are demand for mobility which can be satisfied by bikes. Then, bikes as well as free bike racks stand for the according supply. BSSs can be seen as markets for mobility. We denote a BSS or a station to be balanced if bikes and free bike racks are available when requested. On the other hand side, if requests fail due to the unavailability of bikes or a free bike racks, we denote the BSS or the associated stations to be imbalanced. Then, we define the balancing of BSSs to cover all managerial efforts aiming on the provision of both bikes and free bike racks whenever and wherever requested.

The balancing of BSSs provokes a number of logistical challenges. The challenges are tackled on different management layers. Crainic and Laporte (1997) propose a classification of logistical challenges in freight transportation which has been adapted to logistical challenges in BSSs by Vogel (2016). The management layers differ regarding the horizon of the decisions: strategical (long-term, Sect. 2.5.1), tactical (mid-term, Sect. 2.5.2), and operational management (short-term, Sect. 2.5.3). On every management layer, manifold efforts are spent to support the balancing of BSSs. Figure 2.2 provides an overview. In the following sections, we introduce a number of challenges and associated studies assigned to the respective management layers.

2.5.1 Strategical Management

The strategical management takes decisions with a long-term impact on the BSSs. On this management layer, mainly the BSS's network of stations is designed.

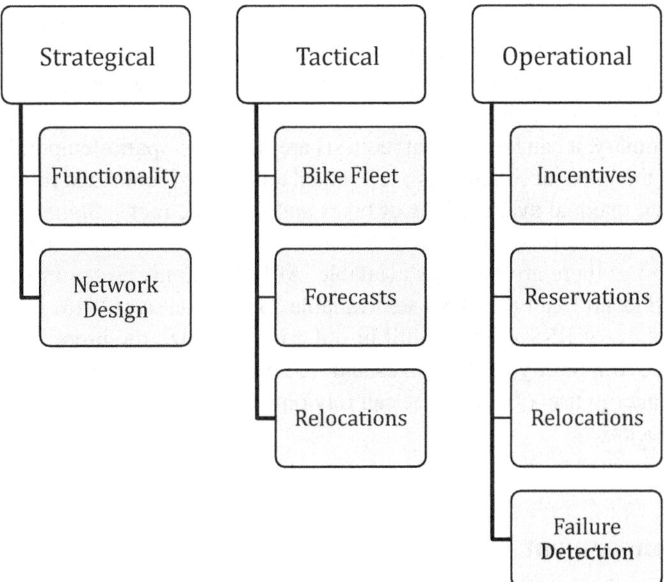

Fig. 2.2 Overview on the management layers

2.5.1.1 Functionality

The very first decision that need to be taken is whether to setup a free-floating or a station-based BSS. Actually, this is rather a political issue than a combinatorial optimization problem (Büttner et al. 2011).

2.5.1.2 Network Design

The network design includes decisions about the number, locations, and capacities of stations. Therefore, the references in this section care about station-based BSSs. The common feature of the following studies is that a number of station locations is selected out of a set of discrete candidate locations. The general assumption is that requests occur at some spatial point. Every station covers a certain area around its location. A request can be served if a station covers the request's spatial point. Otherwise, the request fails as the user is not willing to walk to a faraway station. Therefore, the strategical management aims on providing a network of stations to serve as many requests as possible. Stations with a suitable capacity should be placed at locations where the actual mobility demand occurs. However, each of the following studies comprises unique features.

Lin and Yang (2011) draw on stochastic trips each represented as an origin and a destination. The associated requests occur at the nearest stations. In the approach, the authors decide about station locations and bike lanes. Every bike lane connects two stations. Stations and lanes result in building costs. Every failed request leads to penalty costs whereas served requests lead to travel costs. The travel costs are low if a lane between the user's origin and destination stations is provided. The objective is to minimize the sum over all costs.

García-Palomares et al. (2012) estimate the expected requests based on the residential and working population in certain areas. They decide about station locations as well as capacities. The objective is to serve as many requests as possible subject to a maximum number of stations. The maximum number is exogenously given. The authors evaluate their approach in a case study based on real-world data by the BSS of Madrid (Spain).

Kloimüllner and Raidl (2017b) use expected requests to determine station locations and capacities. Costs occur when stations are build and depend on the capacities. Additional costs due to manual relocations of bikes between stations are estimated and taken into account (see Sect. 3.2). The total costs must not exceed a given budged. The objective is to satisfy as many requests as possible. An evaluation bases on data by the BSS in Vienna (Austria).

Park and Sohn (2017) estimate requests on basis of trips in taxi caps in Seoul (Republic of Korea). They claim that every trip of less than 3 miles ($\approx$4.8 km, Wikipedia 2018) can be replaced by a trip on a bike. Therefore, they decide about station locations and capacities subject to a maximum number of stations. The objective is to minimize the users' travel distances. That is, the walking distance from the origin to a rental station, the cycle distance to the return station, and the walking distance to the actual destination.

Çelebi et al. (2018) decide about both station locations and capacities. A maximum number of bikes must not be exceeded. Stochastic trips are estimated on basis of data by the BSS in Istanbul (Turkey). Therefore, the authors aim on enhancing the BSS. The objective is to maximize the amount of served requests.

2.5.2 *Tactical Management*

The tactical management takes decisions affecting the BSS on a mid-term horizon. That is, the determination of the bike fleet's size, applying information systems forecasting the BSSs development, and optimization regarding the manual relocationing of bikes.

2.5.2.1 Bike Fleet

The size of the bike fleet, i.e., the number of bikes within the BSS has an impact on the availability of both bikes and free bike racks. If the BSS has too many bikes, bike racks will be blocked more often. If the BSS has too little bikes, bikes will be less available. To the best of our knowledge, this problem has not been addressed in the literature yet. However, in station-based BSS, a common practice is to set the number of bikes equal to 50% of the number of bikes racks over all stations (O'Brien et al. 2014).

2.5.2.2 Forecasts

Kaltenbrunner et al. (2010) analyze data by the BSS in Barcelona (Spain). Their data set comprises information about the spatial locations and capacities of stations. Additionally, the respective fill levels in time steps of 2 min are recorded. On basis of this information, the developments of fill levels in the course of the day are extrapolated. Here, the interactions between neighboring stations are taken into account.

Borgnat et al. (2011) predict trips by means of multivariate linear regression. They determine the number of daily trips depending on the members of and bikes in the BSS, the temperature and the amount of rainfall in the respective operating area, and on whether there are holidays or a strike in the respective city. The model is trained with data by the BSS of Lyon (France).

Vogel et al. (2017) introduce an information system to generate artificial trips reflecting the usage pattern of the respective BSS. They draw on data from the BSS in Vienna (Austria). They carry out a cluster analysis (as in their prior work, Vogel et al. 2011) and then study the trips within the clusters as well as the interactions between the clusters. More precisely, for every hour of the day and for every cluster, the number of rentals is determined. Then, for every rental, the return request's cluster is forecasted. Vogel et al. (2017) generalize their findings to create scalable artificial BSSs.

2.5.2.3 Manual Relocations

More research has been done regarding the manual relocationing of bikes by means of transport vehicles. The tactical management draws on expected requests and determines suitable fill levels as well as vehicle tours for a time period like one day or one week in advance. Consequentially, inventory routing problems arise. From the tactical view, the goal is to determine and/or to realize target fill levels and to determine vehicle tours. The objective is to minimize the failed requests when determining the target fill levels. When realizing the target fill levels, the objective is to minimize the tour lengths or the resource consumption in general. Section 3.2 is exclusively dedicated to the class of optimization problems on manual relocationing and comprises a literature classification.

2.5.3 *Operational Management*

A short-term planning horizon is considered by the operational management. That is to use incentives to shift requests, allow reservations, apply manual relocations, and to detect failures of bike and stations.

2.5.3.1 Incentives

In some BSSs, users are involved in the balancing process by means of incentives. The idea is to shift requests from balanced to imbalanced stations. For instance, users of the BSSs in Paris (France) and New York City (NY, USA) are granted extra time for their future trips if they rent bikes at full stations or return bikes at empty stations, respectively (DeMaio 2009; Motivate International Inc. 2018). Compared to manual relocations, user relocations by means of incentives are cheap but less efficient in terms of time (Reiss and Bogenberger 2016).

In traditional economics, a market balance is achieved by means of pricing (Sieg 2008). Therefore, Singla et al. (2015) introduce financial incentives for users if they return their bike at empty stations. When a user requests to rent a bike, he is asked about his destination. If some neighboring stations are empty or have an low fill level, one of these stations is suggested as an alternative for returning the rented bike. Here, the system takes historical data into account to forecast the fill level developments. The user's detour and a price reduction is shown. If the alternative destination station is accepted, the price reduction for the trip is granted. The incentive scheme has been implemented by the BSS in Mainz (Germany) for a field experiment. The results point out, that user are willing to accept the offer with a percentage of 90% if the resulting detour has a length of at most 100 meters, and with a percentages of 50% if the detour has a length of at most 500 meters. These findings are used for experiments combining their incentive scheme with manual relocationing. In the experiments, an optimal split of a certain budget for incentives and additionally manual relocations is identified. For an evaluation, Singla et al. (2015) draw on data by the BSS of Boston (MA, USA). They conclude, that the amount of failed requests can be significantly reduced if the budget is well split.

Reiss and Bogenberger (2016) also study the combination of financial incentives and manual relocationing by means of computational experiments basing on data by the free-floating BSS from Munich (Germany). Here, the operating area is subdivided into zones. For every zone, the imbalances are determined with the assistance of historical data. If the imbalances are high, manual relocationing is applied. Else, the approach draws on user for balancing the associated zone. When a user requests to rent a bike, the smartphone app asks for the destination zone. If the destination zone stated by the user does not have a shortage, an alternative destination and a price reduction are suggested. The alternative destination is a zone with a shortage of bikes. The price reduction depends on the user's detour.

2.5.3.2 Reservations

Kaspi et al. (2014) study the impact of reservation policies for the BSS in Tel Aviv (Israel) by means of a simulation. To avoid failing return requests in the presence of full stations, users are allowed to request bike racks by stating their destination station at their origin station. Then, the reservation policy accepts the reservation if at least one bike rack is free at the destination station. A reserved bike rack is unavailable for other users and, therefore, may also result in failing requests. Nevertheless, according to the simulation, the reservation policy leads to a significant reduction of failed requests compared to the case when a priori requests are not allowed.

2.5.3.3 Manual Relocations

Manual relocationing and determining vehicle tours is addressed also by the operational management. In contrast to the tactical management, the operational aims on explicit short-term decision making with minimum preplanning to encounter unexpected requests and sudden imbalances. From an operational view, the goal is to serve as many requests as possible. Resources such as vehicles are paid and, therefore, vehicle tour lengths can be neglected. Again, we like to refer to Sect. 3.2.

2.5.3.4 Failure Detection

Further, Kaspi et al. (2016) introduce an information system detecting failures of bikes and bike racks. They approximate the probability if a bike is unusable. A bike is most likely unusable if it has not been rented for a long time. Additionally, a return immediately after the rental also indicates some failure. The authors claim that once a failure has been identified, the associated bike can be replaced when relocations are conducted. A data set by the BSS in New York City (NY, USA) is modified for a case study. To this end, bikes are randomly set to be unusable and, therefore, become unavailable for rental requests. According to experiments, the information system reliably identifies unusable bikes.

Chapter 3
Optimization Problems

In this chapter, we highlight the literature related to the balancing of BSSs by means of manual relocationing. In Sect. 3.1, we introduce different types of vehicle routing problems addressing various applications. Section 3.2 focuses on inventory routing problems for BSSs and provides a comprehensive literature classification.

3.1 Vehicle Routing

In this section, we introduce a class of optimization problems where vehicles are dispatched in order to visit customers. We show in which way these problems are related to the balancing of BSSs.

3.1.1 Traveling Salesman Problem

From our point of view, the traveling salesman problem (TSP, Miller et al. 1960) is the most basic variant of vehicle routing. Here, a number of customers needs to be visited by one vehicle. The customers are given by their locations. Hence, the distances between the pairs of customers are known. The vehicle start and end its tour at an arbitrary customer. It is undefined whether the vehicle picks up or delivers commodities. The objective is to schedule a tour with minimal length so that every customer is visited once. The vehicle routing problem (VRP, Toth and Vigo 2014) is a generalization of the TSP where a fleet of vehicles needs to be routed. In many VRPs, the vehicles start and end their tours at a common depot. Solving the TSP to optimality is not possible in polynomial time (Rosenkrantz et al. 1977). Therefore, identifying an optimal solution for the VRP as well as the following extensions also requires non-polynomial time.

The TSP and the VRP are related to the balancing of BSSs as customers can be seen as a correspondence to stations.

J. Brinkmann, *Active Balancing of Bike Sharing Systems*, Lecture Notes in Mobility,
https://doi.org/10.1007/978-3-030-35012-3_3

3.1.2 Capacitated Vehicle Routing Problem

In the capacitated VRP (Laporte et al. 1986), the vehicles underlay capacity restrictions. Likely, the commodities or parcels to be hauled exceed the vehicle capacities. Thus, frequent depot returns are necessary.

In BSSs, vehicle capacities also lead to challenges. For instance, a pickup station needs to be visited if the vehicle is empty, or a delivery station needs to be scheduled next if the vehicle is full. It may also happen that in the presence of an empty or full vehicle only delivery stations, or pickup stations, respectively, are available. In such a case, bikes have to be picked up at or delivered to balanced stations. Then, the challenge is not to initiate failed requests at the balanced station, or, in other words, to *imbalance* the station.

3.1.3 Vehicle Routing Problem with Time Windows

The VRP with time windows (Solomon 1987) adds a time dimension to the VRP. For every customer, an earliest and a latest point in time is given. These two points define a time window in which a vehicle needs to serve the customer. If the time windows must strictly be satisfied, the objective remains to minimize the tour lengths. Otherwise, the objective becomes the minimization of time window violations, i.e., the time the customer is visited too early or too late, respectively.

Time windows play a major role in balancing BSSs since stations must not be visited to early or too late due to the request pattern. If one delivers bikes to a station too early, the station may congest and return requests fail. If one delivers bikes too late, the station may run out of bikes and rental requests fail.

3.1.4 Pickup-and-Delivery Problem

In the pickup-and-delivery problem (Savelsbergh and Sol 1995), parcels have to be hauled. For every parcel, a pickup and a delivery location are given. It is in the nature of things that the pickup location must be visited first and that the delivery locations must be visited by the same vehicle.

With respect to the current fill levels and future requests, BSS stations can be classified to be balanced or to be imbalanced. If station is imbalanced, it is either a pickup or a delivery stations. Notably, we do not have restrictions concerning affiliations of bikes to stations.

3.1.5 Inventory Routing Problem

Inventory routing problems (IRPs) combine characteristics of the aforementioned VRP variants. In IRPs, the goal is to provide customers with a certain commodity

(Coelho et al. 2014b). Every customer has an inventory of the commodity and a consumption ratio. A customer needs to be visited before the corresponding inventory becomes void to avoid penalty costs. Since vehicles have a limited capacity, depot returns are frequently necessary. Customers should not be visited too early if the inventory holding causes costs. Additional costs occur due to the vehicle tour lengths. The objective is to minimize the costs.

The consumption ratio can be seen as an analogy to the request pattern in BSSs. Further, as users request to rent and return bikes, the inventory can be seen as both the stations' bikes and free bike racks. Therefore, every station has two interdependent resources.

Nevertheless, the manual relocationing in BSSs includes a number of additional challenges. Therefore, we discuss inventory routing for BSSs in the next section.

3.2 Inventory Routing for Bike Sharing Systems

In BSSs, bikes as well as free bike racks are resources to satisfy requests. Since the number of bike racks is limited at every station, bikes and free bike racks are interdependent. Additionally, the number of bikes within the system is limited. Generally, the depot does not serve as an unlimited source or sink for bikes. Thus, pickups and deliveries are interdependent. That means if we decide to deliver bikes to some station, we have to pick them up beforehand. IRPs for BSSs consider these extra challenges. In this section, we introduce the related literature.

We provide a classification of literature related to the balancing of station-based BSSs. We distinguish references concerning the modeling of request, the process of decision making, and the objective. Tables 3.1 and 3.2 comprise 57 classified articles. We start with the specification of features and then introduce the categories of the classification. A checkmark ("✓") in a cell refers to an article implementing the corresponding feature. A dash ("–") means that the feature is not implemented. Some features are not applicable ("n/a") if they depend on a certain other feature which is not implemented.

Requests can be modeled stochastically and/or continuously. A checkmark in the column "Stochastic" indicates requests drawn from a stochastic distribution. If requests are modeled stochastically, decisions are taken under uncertainty. If there is a dash in the corresponding column, requests are modeled deterministically and, therefore, are known in advance. Further, requests can occur continuously. "Continuous" requests can occur at any point in time of the considered time horizon. In some articles, the time horizon is divided into periods. Here, all requests of a period occur as a discrete event in the beginning or the ending of the associated period. If no requests are considered, the two request characteristics are not applicable.

If stochastic requests occur, the decision making can either be dynamic or static. "Dynamic" decision are taken in multiple decision points in the course of the time horizon.[1] This is beneficial if unexpected requests occur and the vehicles need to be rescheduled. “Anticipating” indicates the usage of stochastic information when decisions are taken. If either no or deterministic requests occur, dynamic and anticipatory decision making are not applicable. “Active” is also enabled in dynamic decision making only. When active decision making is applied, in one decision point as few decisions as possible are taken and, therefore, the number of decision points is maximized for a certain time horizon. Hence, unexpected requests are counteracted by means of frequent replanning. This is beneficial in the presence of short travel times and a high request frequency (Ulmer et al. 2017a) and enables applicability for the operational management. “Multi-vehicle” decision making involves rebalancing by means of a fleet of multiple vehicles. Decisions to be made concern the “Inventory” of the stations and the “Routing” of the vehicles.[2]

We refer to different objectives with characters (in brackets). The objectives are the minimization either of the vehicle tour lengths (t), the conducted relocations (r), the gaps between target fill levels and realized fill levels (g), or the amounts of failed requests (f). Sometimes, the objectives are indirectly tackled. The vehicle tour lengths can be minimized by considering the total travel times or the exhausted emissions, or by minimizing the waiting times until the stations are served. The gaps can be minimized by maximizing the number of stations served by the vehicles. Failed requests can be translated to costs that are minimized. In some articles, the objective is a combination of the named terms.

We found two main categories where requests are either neglected (Sect. 3.2.1) or considered (Sect. 3.2.2). We describe the categories and show how the categories differ and how the articles in one category are similar to each other. Additionally, we briefly introduce noteworthy articles.

3.2.1 No Request

The first category covers 29 articles (Table 3.1). Here, relocations are usually conducted over night when the system is closed. Thus, requests are not considered. The goal is to schedule tours and relocation operations to realize target fill levels. The target fill levels are provided by external information systems and can be seen as initial fill levels when the system is opened in the next morning. Therefore, the models usually do not comprise a time dimension.[3] Consequentially, it is irrelevant whether a certain relocation is conducted in the beginning of the time horizon or in the end.

---

[1] We want to note that also in deterministic-static optimization problems multiple decision points are sometimes used. This happens due to decomposition.

[2] Strictly speaking, if either no inventory or no routing decisions is made, no IRP is investigated in the associated article. Nevertheless, one of these decisions may not be included due to decomposition. We add the articles to our classification since they are strongly related to IRPs for BSSs.

[3] Of course, the vehicle tour lengths can be translated to total travel times.

In most articles, inventory as well as routing decisions are made. The objectives are to minimize the tour length and/or the gaps between targets and realized fill levels.

From our point of view, Chemla et al. (2013) introduce the most basic IRP for BSSs. One vehicle is dispatched to pick up and deliver bikes. All given target fill levels need to be realized. The objective is to determine tours of minimal length. Kloimüllner et al. (2015) and Kloimüllner and Raidl (2017a) introduce a multi-vehicle model where no inventory decisions need to be made. Here, stations are beforehand classified as pickup and delivery stations. The shortage of either bikes or free bike racks is not specified. The objective is to schedule minimal tours being subject to alternating visits of pickup and delivery stations. In the model introduced by Kadri et al. (2016, 2018), a target fill level for every station is given. Then, the objective is to minimize the total time until every stations' fill level matches the target. Schuijbroek et al. (2017) introduce a multi-vehicle model. Bikes are relocated with respect to target fill levels (which notably are determined by a preceding information system introduced in the same article) at rush hours. The authors state that requests during the relocation process can be neglected since they are incorporated in the target fill levels. Wang and Szeto (2018) draw on an approach determining and minimizing the exhausted emissions.

3.2.2 *Request*

In the second category, including 28 articles (Table 3.2), relocations are conducted during the day when the system is opened. Thus, requests take place when the vehicles are on the road. As mentioned in Sect. 2.5.2, one can draw on external information systems extracting requests from real-world data.

This category can further be subdivided into the three subcategories concerning the representation of requests and the decision making process. In Sect. 3.2.2.1, requests are deterministic and, therefore, known when decisions are taken. As an inference, the decision making is static. If requests are stochastic and, therefore, uncertain when decisions are taken, the decision making process can either be static or dynamic. Decision making is static if decisions are taken in one decision point. In contrast, decision making is dynamic if decisions are taken over a sequence of decision points. In Sect. 3.2.2.2, we introduce the subcategory on stochastic-static IRPs for BSSs. The subcategory on stochastic-dynamic approaches is introduced in Sect. 3.2.2.3.

3.2.2.1 Deterministic-Static Models

In the first subcategory, comprising 12 articles, requests are deterministic and decisions making is static. Since requests are considered, the minimization of failed requests is an additional and potential objective.

Caggiani and Ottomanelli (2012, 2013) determine relocations conducted by one vehicle. The vehicle serves the stations in a sequence corresponding to an TSP tour.

Table 3.1 Literature classification on inventory routing for bike sharing systems: no requests

References	Requests		Decision Making						Objective
	Stochastic	Continuous	Dynamic	Anticipating	Active	Multi-Vehicle	Inventory	Routing	
Chemla et al. (2013)	n/a	n/a	n/a	n/a	n/a	–	✓	✓	t
Di Gaspero et al. (2013a)	n/a	n/a	n/a	n/a	n/a	✓	✓	✓	g, r, t
Di Gaspero et al. (2013b)	n/a	n/a	n/a	n/a	n/a	✓	✓	✓	g, t
Papazek et al. (2013, 2014)	n/a	n/a	n/a	n/a	n/a	✓	✓	✓	g, r, t
Rainer-Harbach et al. (2013, 2015)	n/a	n/a	n/a	n/a	n/a	✓	✓	✓	g, r, t
Raviv et al. (2013)	n/a	n/a	n/a	n/a	n/a	✓	✓	✓	g, t
Dell'Amico et al. (2014, 2016)	n/a	n/a	n/a	n/a	n/a	✓	–	✓	t
Erdoğan et al. (2014)	n/a	n/a	n/a	n/a	n/a	–	✓	✓	t, r
Ho and Szeto (2014)	n/a	n/a	n/a	n/a	n/a	–	✓	✓	g
Erdoğan et al. (2015)	n/a	n/a	n/a	n/a	n/a	–	✓	✓	t
Forma et al. (2015)	n/a	n/a	n/a	n/a	n/a	✓	✓	✓	g, t
Kloimüllner et al. (2015)	n/a	n/a	n/a	n/a	n/a	✓	–	✓	g
Alvarez-Valdes et al. (2016)	n/a	n/a	n/a	n/a	n/a	✓	✓	✓	g
Espegren et al. (2016)	n/a	n/a	n/a	n/a	n/a	✓	✓	✓	g, r, t
Kadri et al. (2016)	n/a	n/a	n/a	n/a	n/a	–	✓	✓	g
Li et al. (2016)	n/a	n/a	n/a	n/a	n/a	–	✓	✓	g, t

(continued)

Table 3.1 (continued)

References	Requests		Decision Making						Objective
	Stochastic	Continuous	Dynamic	Anticipating	Active	Multi-Vehicle	Inventory	Routing	
Szeto et al. (2016)	n/a	n/a	n/a	n/a	n/a	–	✓	✓	g, t
Cruz et al. (2017)	n/a	n/a	n/a	n/a	n/a	–	✓	✓	t
Ho and Szeto (2017)	n/a	n/a	n/a	n/a	n/a	✓	✓	✓	g, t
Kloimüllner and Raidl (2017a)	n/a	n/a	n/a	n/a	n/a	✓	–	✓	g
Schuijbroek et al. (2017)	n/a	n/a	n/a	n/a	n/a	✓	✓	✓	t
Arabzad et al. (2018)	n/a	n/a	n/a	n/a	n/a	✓	✓	✓	t
Bulhões et al. (2018)	n/a	n/a	n/a	n/a	n/a	✓	✓	✓	t
Kadri et al. (2018)	n/a	n/a	n/a	n/a	n/a	✓	✓	✓	g
Szeto and Shui (2018)	n/a	n/a	n/a	n/a	n/a	✓	✓	✓	g, r, t
Wang and Szeto (2018)	n/a	n/a	n/a	n/a	n/a	✓	✓	✓	t

Table 3.2 Literature classification on inventory routing for bike sharing systems: requests

References	Requests		Decision Making						Objective
	Stochastic	Continuous	Dynamic	Anticipating	Active	Multi-Vehicle	Inventory	Routing	
Caggiani and Ottomanelli (2012, 2013)	–	✓	n/a	n/a	n/a	–	✓	–	r, t
Contardo et al. (2012)	–	✓	n/a	n/a	n/a	✓	✓	✓	t
Kloimüllner et al. (2014)	–	✓	n/a	n/a	n/a	✓	✓	✓	f, g, r, t
Vogel et al. (2014, 2017)	–	–	n/a	n/a	n/a	✓	✓	–	f, t
Neumann Saavedra et al. (2015)	–	–	n/a	n/a	n/a	✓	✓	–	f, t
Brinkmann et al. (2016)	–	–	n/a	n/a	n/a	✓	✓	✓	g
Neumann Saavedra et al. (2016)	–	–	n/a	n/a	n/a	✓	✓	✓	f, t
Shui and Szeto (2017)	–	–	n/a	n/a	n/a	–	✓	✓	f, t
Zhang et al. (2017)	–	–	n/a	n/a	n/a	✓	✓	✓	f, t
Tang et al. (2019)	–	–	n/a	n/a	n/a	✓	✓	✓	f, t
Raviv and Kolka (2013)	✓	✓	–	✓	n/a	–	✓	–	f
Kadri et al. (2015)	✓	✓	–	✓	n/a	–	✓	✓	g
Ghosh et al. (2016)	✓	–	–	✓	n/a	✓	✓	✓	f
Lu (2016)	✓	–	–	✓	n/a	✓	✓	–	f, t
Datner et al. (2017)	✓	✓	–	✓	n/a	–	✓	–	f
Ghosh et al. (2017)	✓	–	–	✓	n/a	✓	✓	✓	f, t
Yan et al. (2017)	✓	–	–	✓	n/a	✓	✓	–	f, t
Dell'Amico et al. (2018)	✓	✓	–	✓	n/a	✓	✓	✓	f, t
Neumann Saavedra (2018)	✓	✓	–	✓	n/a	✓	✓	✓	f, t
Brinkmann et al. (2015)	✓	✓	✓	–	✓	–	✓	✓	f

(continued)

Table 3.2 (continued)

References	Requests		Decision Making						Objective
	Stochastic	Continuous	Dynamic	Anticipating	Active	Multi-Vehicle	Inventory	Routing	
Ricker (2015)	✓	✓	✓	✓	–	✓	✓	✓	f, t
Fricker and Gast (2016)	✓	✓	✓	–	✓	–	✓	✓	g
Brinkmann et al. (2019a)	✓	✓	✓	✓	✓	–	✓	✓	f
Brinkmann et al. (2019b)	✓	✓	✓	✓	✓	✓	✓	✓	f
Chiariotti et al. (2018)	✓	–	✓	✓	–	–	✓	✓	r, t
Legros (2019)	✓	✓	✓	✓	–	–	✓	✓	f

Vogel et al. (2014, 2017) and Neumann Saavedra et al. (2015) determine relocations that comprise a pickup station, a delivery station, the number of bikes to be relocated, and a time period when the relocations should be conducted. The relocations are isolated, i.e., vehicle tours are not determined. The determined relocations can be accumulated to achieve time-dependent target fill levels. The authors assume that every relocation and every failed request leads to specific costs. The objective is to minimize the costs over all relocations and failed requests. Brinkmann et al. (2016) implement the target fill levels provided by Vogel et al. (2014). The target fill levels are realized in a multi-vehicle model. The objective is to minimize the squared gaps between realized fill levels and target fill levels subject to a limited time horizon. The authors show that predetermined target fill levels can be realized in a deterministic-static environment. Shui and Szeto (2017) also draw on target fill levels and minimize a combined term including the failed requests and the exhausted emissions.

3.2.2.2 Stochastic-Static Models

We identify 9 articles for the second subcategory. Here, requests are modeled stochastically and decision making is static.

Raviv and Kolka (2013) determine target fill levels for isolated stations. Since the requests at neighboring stations are interdependent (Rudloff and Lackner 2014), Datner et al. (2017) determine target fill levels for all stations of a BSSs. The target fill levels by Raviv and Kolka (2013) as well as Datner et al. (2017) are determined on stochastically modeled requests and serve as initial fill levels for the morning when the systems is opened. The authors assume that these fill levels are realized during the night and that no additional relocations are conducted during the day. Therefore, the required input for the articles of the first category (Sect. 3.2.1) is provided. The objective considered by Kadri et al. (2015) is to minimize the total times the stations are empty or full. Ghosh et al. (2016, 2017), Lu (2016) and Yan et al. (2017) model stochasticity by using multiple realizations of requests with probabilities of occurrences. The authors subdivide the time horizon into disjoint periods and aggregate requests for every period. Anticipation is enabled by considering the probabilities of occurrences when taking decisions. Lu (2016) and Yan et al. (2017) make inventory decisions only as vehicle tours are given. Neumann Saavedra (2018) draws on prior work and introduces an evaluation on stochastic and continuous requests.

3.2.2.3 Stochastic-Dynamic Models

The third subcategory comprises articles with stochastic requests and dynamic decision making (7 articles). In all articles, requests are modeled stochastically.

Fricker and Gast (2016), Brinkmann et al. (2015, 2019a), Chiariotti et al. (2018) and Legros (2019) dispatch one vehicle. Fricker and Gast (2016) dispatch the vehicle such that it picks up bikes at the station with the highest fill levels and delivers bikes to the station with the lowest fill level. Brinkmann et al. (2015) introduce safety buffers, i.e., a minimum number of bikes and free bike racks at every station. The vehicle is sent to its nearest station where the safety buffers are violated. Brinkmann et al. (2019a) draw on resampled requests and simulations to anticipate failing requests and to evaluate potential inventory decisions. Brinkmann et al. (2019b) extend their prior work and introduce a multi-vehicle model. The vehicles are dispatched by means of solving assignment problems. Ricker (2015), Chiariotti et al. (2018) and Legros (2019) introduce models where target fill levels are determined for certain periods. Also the decisions are taken for every period at once. Chiariotti et al. (2018) models aggregated requests that also occur at once as a stochastic event in every period. Ricker (2015) schedules multiple vehicles.

In this work, we draw on the model by Brinkmann et al. (2019b) as it is the most comprehensive one. To this end, we have to establish a basis on dynamic decision making in the next chapter.

Chapter 4
Dynamic Decision Making

In this chapter, we define the basic concepts on dynamic decision making. We first introduce the concept of a dynamic decision process (DDP). In Sect. 4.1, we define the Markov decision process as a special case of DDPs and as the modeling technique. In Sect. 4.2, we introduce approximate dynamic programming as a collection of policies to solve MDPs.

In an DDP, a number of decisions is taken in a defined system over a time horizon. Figure 4.1 depicts an DDP and its elements. A decision point k is induced when a predefined event occurs. An exogenous process continuously influences the system. The process is exogenous in the way that the decision maker does not control it. The system is defined by a set of parameters. A decision state $s_k \in S$ refers to the system parameters' characteristics in decision point k. Additionally, s_k defines a set of feasible decisions X_{s_k}. A decision $x_k \in X_{s_k}$ needs to be selected. The decision leads to a contribution $p_k \in \mathbb{R}$. The contribution might be a reward which needs to be maximized, or a penalty which needs to be minimized.[1] Furthermore, by selecting (and realizing) a decision, the decision maker influences the system. Therefore, the subsequent state s_{k+1} bases on the previous state s_k, on the selected decision x_k as well as on the exogenous process.

4.1 Markov Decision Processes

A Markov decision process, as defined by Puterman (2014), is a special case of the DDP. In an MDP, the decision maker exclusively draws on the current state parameters when selecting a decision. Additionally, the exogenous process is uncertain. Likely, it is subject to a stochastic distribution.

The elements of an MDP are depicted by Fig. 4.2. When a decision point k occurs, the associated decision state $s_k \in S$ is revealed. Then, a decision $x \in X_{s_k}$ is selected.

[1]If not stated otherwise, we assume a contribution to be a penalty.

J. Brinkmann, *Active Balancing of Bike Sharing Systems*, Lecture Notes in Mobility,
https://doi.org/10.1007/978-3-030-35012-3_4

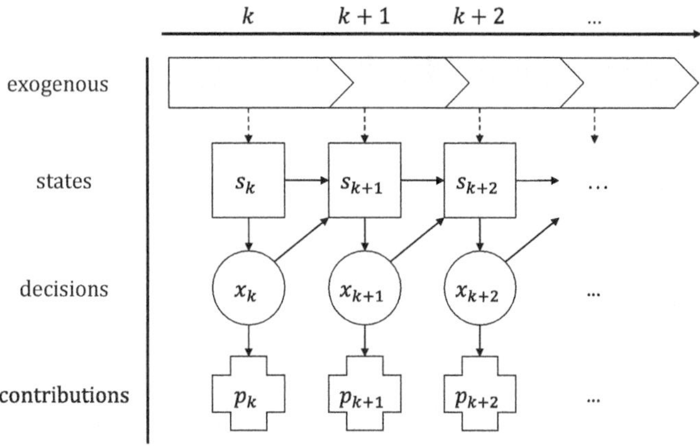

Fig. 4.1 A dynamic decision process (adapted, Meisel 2011)

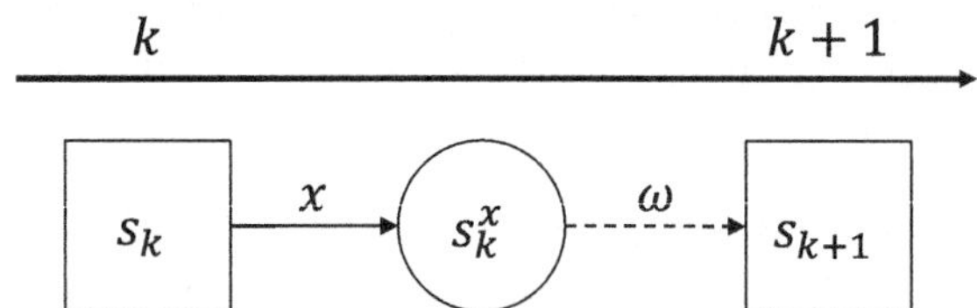

Fig. 4.2 A Markov decision process (adapted, Ulmer et al. 2015)

The combination of decision state and decision leads to a deterministic post-decision state $s_k^x = (s_k, x)$. An associated penalty might deterministically depend on s_k and x: $p(s_k, x)$. Otherwise, it additionally depends on the realization of the exogenous process $\omega \in \Omega$ and, therefore, is stochastic: $p_\omega(s_k, x)$. When the decision is selected, the exogenous process between two decision points is revealed. It alters the corresponding state parameters. Therefore, we also denote ω as transition function. Then, $\omega(s_k, x) = s_{k+1}$ is the subsequent decision state.

A solution of an MDP is a policy $\pi \in \Pi$. For every decision state $s_k \in S$, a policy $\pi : S \rightarrow X$ returns a decision $x \in X_{s_k}$.[2] Here, p_ω represents the penalties with respect to ω. Thus, p_ω is uncertain until ω is revealed. As depicted in Eq. (4.1), the objective is to identify an optimal policy $\pi^* \in \Pi$ minimizing the expected sum of penalties over all decision points conditioned on the initial decision state s_0:

$$\pi^* = \arg\min_{\pi \in \Pi} \mathbb{E}\left[\sum_{k=0}^{k_{\max}} p_\omega\big(s_k, \pi(s_k)\big)\Big| s_0\right] \tag{4.1}$$

An optimal policy π^* selects decisions using the Bellman Equation (4.2, Bellman 1957).

[2]For the sake of simplicity, we denote the super set over all decision sets as $X = \cup_{s \in S} X_s$.

$$\pi^*(s_k) = \underset{x \in X_{s_k}}{\arg\min}\, p(s_k, x) + \underbrace{\mathbb{E}\left[\sum_{k'=k+1}^{k_{\max}} p_\omega\left(s_{k'}, \pi^*(s_{k'})\right)\middle| s_{k+1}\right]}_{v(s_k, x)} \tag{4.2}$$

As we can see, π^* selects a decision with respect to the penalty associated with k and with the expected sum of penalties associated with all future decision points as well. For the sake of simplicity, we denote the expected sum of future penalties as value. We read value $v(s_k, x) \in \mathbb{R}$ as the expected sum of future penalties if decision x is selected in decision state s_k. Generally speaking, incorporating future events into a decision making process is called anticipation (Butz et al. 2003). Therefore, an optimal policy anticipates the future developments of the MDP. In the MDP, future decision states are influenced by the selected decision and the transition. To determine value $v(s_k, x)$, and in this way, to select the optimal decision, one needs to take all feasible developments of the MDP into account. One can illustrate all feasible developments of an MDP by a decision tree. Figure 4.3 illustrates an exemplary decision tree depicting an MDP's development over decision points k, $k+1$, and $k+2$. Starting in decision state s_k (depicted by a square) on the left-hand side, two decisions $x_0, x_1 \in X_{s_k}$ are feasible. The solid edges depict selections of associated decisions. Therefore, two post-decision states $s_k^{x_0}, s_k^{x_1}$ (depicted by circles) follow. The outgoing dashed edges depict realizations of transitions. Since the transition is stochastic, more subsequent decision states may occur than feasible decisions exist. However, it is possible that different decisions result in the same decision state. In the example, three different decision states may occur in decision point $k+1$. In the step from $k+1$ to $k+2$, the number of possible decision states increases once more due to the combination of decisions and transitions.

Given the transition probabilities are known, we can determine the expected future penalties by means of dynamic programming. In essence, dynamic programming refers to total enumeration of the decision tree. In this way, we can solve any stochastic-dynamic optimization problem to optimality. However, the tree might grow exponentially due to the Curses of Dimensionality (Powell 2011):

The State Space

The number of states $s \in S$ depends on the state attributes and their domains. Let $s = (a_1, \ldots, a_{\max})$ be a decision state defined by parameters $a_i \in A_i$. Additional restrictions may forbid certain combinations of parameters. Therefore, the number of feasible states is at most equal to the cardinality of the Cartesian product of all state parameters' domains: $|S| \leq |A_1 \times \cdots \times A_{\max}|$.

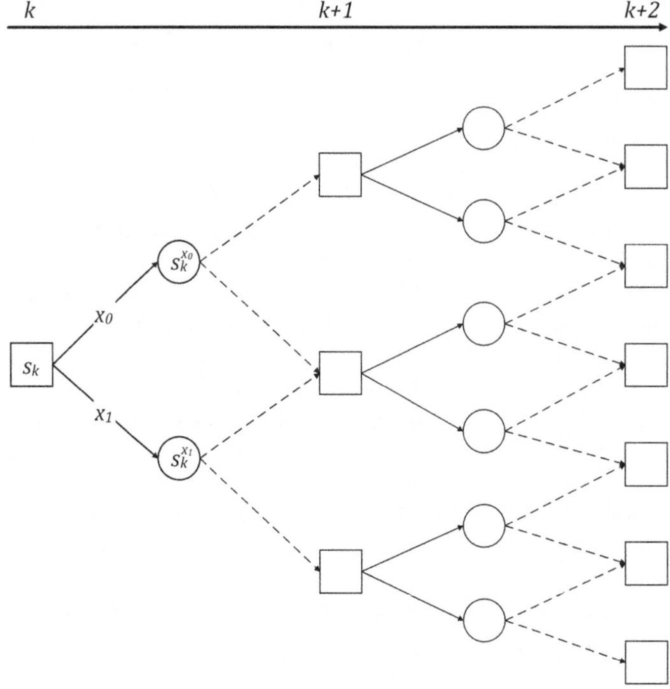

Fig. 4.3 An exemplary decision tree (adapted, Ulmer 2017)

The Solution Space

As stated beforehand, an MDP's solution is a policy. Therefore, the solution space corresponds to the set of policies Π. The policies differ regarding the decisions they yield for the various decision states $s \in S$. According to the objective (4.1), we aim on identifying an optimal policy minimizing the expected penalties with respect to *any* MDP. Therefore, the number of policies is the cardinality of the Cartesian product of *all* decision sets associated with the decision states: $|\Pi| = |X_{s_0} \times \cdots \times X_{s_{\max}}|$.

The Transition Space

After selecting a decision $x \in X_{s_k}$ in decision state s_k, a transition function ω reveals the next decision state: $\omega(s_k, x) = s_{k+1}$. Since in every decision state $s_k \in S$, any decision $x \in X_{s_k}$ can be selected, and any state $s_{k+1} \in S$ may follow, the number of feasible transitions is the cardinality of the Cartesian product of the three components: $|\Omega| = |S \times X \times S| = |S^2 \times X|$.

The state space, the solution space, an the transition space usually grow with exponential expansion rates. Therefore, total enumeration of the decision tree is hardly possible if too many state attributes are considered. Nevertheless, we draw on approximate dynamic programming to identify policies providing both reasonable solution quality in acceptable runtime.

Regarding the runtime, we have to distinguish the runtime in a single decision point and the runtime for the entire MDP. In a real-world application, the runtime in every decision point is crucial if decisions have to be taken on short notice. On the other hand side, the MDP's runtime is neglectable as it is distributed over the decision points. The MDP's runtime is significant when a policy is evaluated in computational experiments. To obtain accurate performance indicator (e.g., the average amount of penalties or the runtime itself), a large number of test instances has to be solved. Further, policies may require parameters to be defined beforehand. This results in additionally experiments. Here, a huge runtime of the MDP may prevent reasonable experiments.

4.2 Approximate Dynamic Programming

Approximate dynamic programming (ADP) covers a broad area of policies to solve stochastic-dynamic optimization problems heuristically. In contrast to dynamic programming aiming on determining the true values and resulting in total enumeration of the decision tree, ADP *approximates* the values (Powell 2011).

In this work, we introduce three categories of policies: Myopic (Sect. 4.2.1), Lookahead (Sect. 4.2.2), and Value Function Approximation (VFA, Sect. 4.2.3). The categories are distinguished by means of four characteristics. An overview is presented in Table 4.1. If the policies of a category enable anticipation, the corresponding cell includes a checkmark ("✓"). In the considered categories, anticipation is realized by means of simulations. The simulations can be conducted "ad hoc", i.e., when the corresponding decision point occurs, or "a priori", i.e., before the actual MDP is solved. The simulation's level of detail is either "high" or "low". If anticipation is not enabled, simulations and the level of detail are not applicable ("n/a"). Every policy comprise an algorithmic procedure conducted when a decision point occurs (e.g., a simulation or a search operation in a data structure). The procedures are either computational expensive or cheap. This results in either "high" or "low" runtime in single

Table 4.1 Overview on categories of approximate dynamic programming

Category	Anticipation	Simulation	Level of detail	Runtime in k
Myopic	–	n/a	n/a	Low
Lookahead	✓	Ad hoc	High	High
VFA	✓	A priori	Low	Low

decision points. Anticipation enabled by high detailed simulations is advantageous in terms of the solution quality. However, the runtime in single decision points is crucial when decision are required on short notice.

We introduce the categories in the following sections in detail. Further, we introduce an exemplary policy for every category. The exemplary policies can be seen as blue prints for optimization in various applications.

4.2.1 Myopic

We define a policy to be myopic if it ignores the future development of the MDP and, therefore, does not enable anticipation. Decisions are usually made based on state parameters and a computationally cheap algorithm. Some myopic policies also have tuning parameters such as safety buffers for inventory levels. Greedy policies are special cases of myopic policies. Decisions are made only with respect to the penalties associated with the decisions in the current decision state[3] shown in Eq. (4.3).

$$\pi_{\text{greedy}}(s_k) = \arg\min_{x \in X_{s_k}} p(s_k, x) \tag{4.3}$$

The strength of myopic policies is the ability to return decisions almost without any delay. This is vital in the context of real-time decision making. The weakness is that the future development of the MDP and associated penalties are ignored. Therefore, the solution quality is often low.

4.2.2 Lookahead

Lookahead policies take future developments of the MDP into account when selecting a decision. The impact of a decision is approximated by means of simulations of the MDP at runtime. Based on these simulations, the expected future penalties associated with the decisions are approximated and, therefore, anticipation is enabled.

Rollout algorithms, as introduced by Goodson et al. (2017), are special cases of lookahead policies using computationally cheap (and most likely myopic) base policies to simulate future decisions. For any decision $x \in X_{s_k}$, a rollout algorithm simulates feasible decisions to approximate the value. The procedure is depicted by Fig. 4.4. Starting in any post-decision state s_k^x, multiple simulations of the MDP are conducted. Let $\omega_1, \ldots, \omega_{\max} \in \Omega' \subset \Omega$ be resampled realizations of Ω and let π_{base} be a base policy. A rollout algorithm conducts a simulation for every decision and resampled realization. As the MDP itself is simulated, the simulations' level of detail is high. In the simulations, the rollout algorithm rolls the base policy out, i.e.,

[3] In other words, a greedy policy "approximates" every value to be equal to zero.

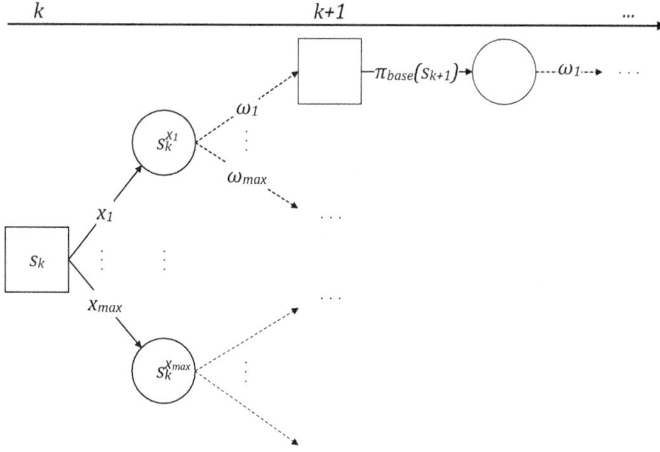

Fig. 4.4 A rollout algorithm's simulations (adapted, Ulmer et al. 2016)

the decisions are made by the base policy. In this way, the rollout approximates the values $v(s_k, x)$ by the average penalties in the simulations. As depicted by Eq. (4.4), in s_k, the rollout selects $x \in X_{s_k}$ according to the penalties associated with x and the average sum of penalties followed by s_k^x over all associated simulations:

$$\pi_{\text{rollout}}(s_k) = \underset{x \in X_{s_k}}{\arg\min}\; p(s_k, x) + \underbrace{\frac{1}{|\Omega'|} \cdot \sum_{\omega \in \Omega'} \sum_{k'=k+1}^{k_{\max}} p_\omega\left(s_{k'}, \pi_{\text{base}}(s_{k'})\right)}_{\text{average penalties over all simulations}} . \tag{4.4}$$

The strength of lookahead policies is the ability to approximate future penalties and, therefore, to anticipate the future development of the MDP. The approximations are precise since the simulations base directly on the MDP. The first weakness is that the exogenous process needs to be represented by a stochastic function. Usually, this requires a huge amount of data to be recorded and preprocessed before any simulation can be conducted. The second weakness is that the simulations are computational expensive. The number of simulations that needs to be conducted before a reliable approximation is achieved differs and need to be determined by means of experiments. Since the simulations are conducted ad hoc when a decision point occurs, a large number of simulations may prevent decision making in certain applications. For rollout algorithms, the base policy is a third factor. A rollout algorithm's performance strongly depends on the decisions made in the simulations. Therefore, a base policy leading to a certain minimum solution quality is required.

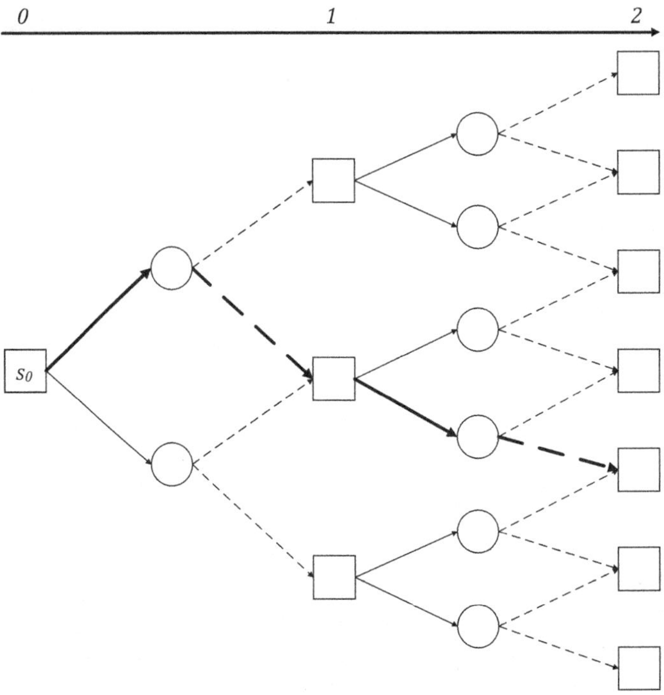

Fig. 4.5 An exemplary trajectory in a decision tree

4.2.3 Value Function Approximation

The idea of a value function approximation (VFA) is to approximate the values a priori in an approximation phase, i.e., before the actual MDP is solved. Anticipation is realized by means of applying the approximated values in an application phase.

In the approximation phase of a non-parametric VFA,[4] we technically aim on approximating a function $\tilde{v}\colon S \times X \rightarrow \mathbb{R}$ where the approximated values are close to the real values: $\tilde{v}(s, x) \approx v(v, x), \forall s \in S, x \in X$. The individual values are iteratively approximated by means of traversing trajectories in the decision tree and by observing the penalties resulting from decisions selected in certain decision states. A trajectory is a realization of an MDP starting in the initial decision state and ending in the final. A trajectory can be depicted as a path in the decision tree. Figure 4.5 shows an exemplary trajectory embedded in a decision tree as discussed in Sect. 4.1. The bold edges depict the trajectory. Starting in the initial decision state s_0, the trajectory extends to a decision state s_2 which is the final decision state in this example. In the

[4]In a parametric VFA, one assumes a functional dependency between state parameters and value. Since we do not apply parametric VFA in this work, we will not go into details. However, we like to refer the interested reader to Powell (2011) for a deeper overview on VFA and to Wooldridge (2015) for an overview on the determination of multi-variate linear functions by means of econometrics.

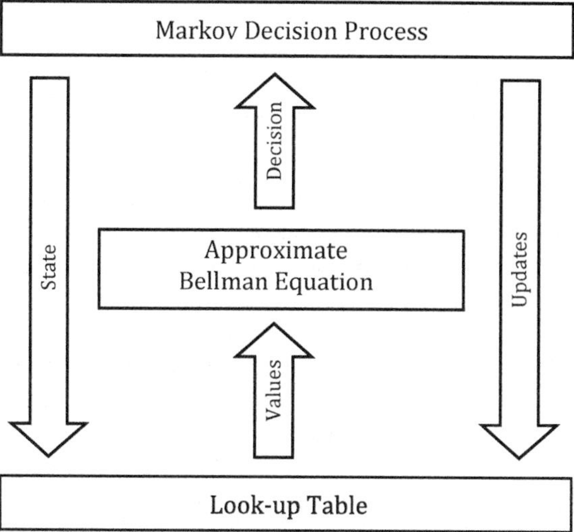

Fig. 4.6 An iteration of an VFA's approximation phase

trajectory's decision points, the decisions are selected with respect to the incomplete approximated values.[5] A resampled realization of the exogenous process yields the transitions. To approximate the values iteratively, we traverse a sequence of trajectories.

In the approximation phase, we store the approximated values in a lookup table (LUT). An LUT is a data structure comprising values for every state and decision. Before the first trajectory is traversed, the values are assigned an initial magnitude. Then, traversing trajectories and approximating values is realized as follows. Figure 4.6 depicts one iteration of the approximation phase. Whenever a decision point k occurs in the current trajectory's MDP, we hand over the decision state s_k to the LUT. The LUT returns the values of all feasible decisions $x \in X_{s_k}$ to the approximate Bellman Equation (4.5).

$$\pi_{\text{VFA}}(s_k) = \underset{x \in X_{s_k}}{\arg\min}\; p(s_k, x) + \tilde{v}(s_k, x) \tag{4.5}$$

Here, the decision $\pi_{\text{VFA}}(s_k)$ with minimal immediate and future penalties is selected and applied in decision point s_k of the trajectory's MDP. When the final decision point $k_{\max}$ is processed, the values of all decision states occurred and decisions selected in the current trajectory are updated. Let k be a decision point and let s_k and x be the associated decision state or decision, respectively. Further, $\alpha_{s_k,x} \in \mathbb{N}_0$ depicts the number of occurrences of a specific decision state s and a decision x over all trajectories traversed for far in the approximation phase. Then, the values are updated according to Eq. (4.6).

[5]The term "incomplete approximated" refers to the fact that the approximation phase is still ongoing.

$$\tilde{v}(s_k, x) := \underbrace{\frac{\alpha_{s_k,x} - 1}{\alpha_{s_k,x}} \cdot \tilde{v}(s_k, x)}_{\text{old approximation}} + \underbrace{\frac{1}{\alpha_{s_k,x}} \cdot \sum_{k'=k+1}^{k_{\max}} p_\omega\big(s_{k'}, \pi_{\text{VFA}}(s_{k'})\big)}_{\text{new observation}} \tag{4.6}$$

The left term depicts the old approximation before the current trajectory. The right term depicts the observed penalties *after* the decision has been selected in the current trajectory. The terms are weighted by the factors including $\alpha_{s_k,x}$ such that the values are equal to the average future penalties over all trajectories.[6]

With ongoing trajectories in the approximation phase and successive updating of values, we obtain more precise approximations of the values. As soon as the values are approximated, the value function can be applied for a further evaluation or in a real-world application by using the approximate Bellman Equation (4.5).

The strength of VFA-based policies is the ability to adapt decision making to the exogenous process. Reliable and beneficial decision making is enabled if the values are approximated well. Further, we do not need an algorithmic optimization at runtime since we can directly access the approximated values in the LUT. The challenges we face are various. First, the exogenous process needs to be modeled by a stochastic function as already discussed in the section on lookahead policies. Second, the approximation phase is time consuming. Therefore, we cannot apply a VFA-based policy ad hoc (in contrast to lookahead policies). Further, a termination criterion for the approximation phase needs to be defined. Third, there may be too many state/decision combinations and, therefore, too many values so that the approximation phase cannot yield accurate approximations in reasonable time. In such a case, a partitioning of the state space is applied to support the approximation phase. That means, similar (but different) states are assigned to one entry in the LUT. Consequentially, the LUT comprises fewer values. In this way, the approximation phase is accelerated. However, a state space partitioning leads to a coarse granularity of the LUT, i.e., some attributes (or attributes' details) are ignored. As a consequence, the solution quality is most likely decreased. One has to find a compromise to solve this tradeoff.

[6]As soon as a specific decision state occurred and a specific decision has been selected, α is incremented. Therefore, α is never equal to zero when Eq. (4.6) is applied.

Part II
Application

Chapter 5
The Stochastic-Dynamic Multi-Vehicle Inventory Routing Problem for Bike Sharing Systems

In this chapter, we introduce the multi-vehicle stochastic-dynamic inventory routing problem for bike sharing systems (IRP_{BSS}) as first presented by Brinkmann et al. (2019b).

We provide a narrative formulation of the IRP_{BSS} in Sect. 5.1. In Sect. 5.2, the infrastructure is formally defined. We model the IRP_{BSS} as an MDP in Sect. 5.3. In Sect. 5.4, an example of the MDP is presented. We point out the resulting challenges when solving the MDP in Sect. 5.5.

5.1 Narrative

We consider a BSS consisting of stations and a depot. Every station has a capacity, i.e., a limited number of bike racks. The fill level of a station denotes the number of available bikes. The difference of capacity and fill level depicts the number of free bike racks. In the BSS, users undertake trips. Each trip consists of at least one rental request and at least one return request which are unknown until they occur. A rental request occurs at the station where a user requests to rent a bike. If a bike is available, the rental request is served and the user cycles to another station to return the bike. Else, the rental requests fails and the user walks to another station to rent a bike. A return request occurs at the station where a user requests to return the bike. If a free bike rack is available, the return request is served and the user returns the bike. Else, the return request fails and the user cycles to another station to return the bike. The provider's goal is to minimize the amount of failed requests. To this end, he draws on manual relocations and applies a fleet of vehicles. In the beginning of the day, the vehicles are located at the depot. Then, the vehicles drive along their tours and visit stations. Whenever a vehicle arrives at a station, the station's inventory is adapted by means of relocations. When the relocations are completed, the vehicle is routed to its next station. The vehicles are also subject to a capacity, i.e., a limited number of bikes they can transport at one point in time. In the end of the day, the vehicles return to the depot.

J. Brinkmann, *Active Balancing of Bike Sharing Systems*, Lecture Notes in Mobility,
https://doi.org/10.1007/978-3-030-35012-3_5

Table 5.1 Notation of the bike sharing system's infrastructure

Symbol	Description
$N = \{n_0, \ldots, n_{\max}\}$	Set of stations
$c_n = (c_{n_0}, \ldots, c_{n_{\max}}) \in \mathbb{N}_0^{\vert N\vert}$	Stations' capacities
$V = \{v_1, \ldots, v_{\max}\}$	Set of vehicles
$c_v = (c_{v_1}, \ldots, c_{v_{\max}}) \in \mathbb{N}^{\vert V\vert}$	Vehicles' capacities
$\tau \in \mathbb{N}$	Travel time between two stations
$\tau_r \in \mathbb{N}$	Service time for one relocation

5.2 Infrastructure

We first introduce the notation of the BSS's infrastructure. Table 5.1 provides an overview about the symbols. The infrastructure is fixed in the sense that it does not change during the day.

The set of stations is denoted by N. Every station $n_i \in N$ has a capacity $c_{n_i} \in \mathbb{N}_0$. In the model, the depot n_0 is a station with capacity $c_{n_0} = 0$.[1] The vehicles are depicted by $v \in V$. Every vehicle v_i has a capacity $c_{v_i} \in \mathbb{N}$. The vehicles' travel time between two stations n_i and n_j is given by $\tau_{n_i,n_j} \in \mathbb{N}$. The relocation of one bike requires a service time of $\tau_r \in \mathbb{N}$.

5.3 Markov Decision Process

The IRP_{BSS} matches the definition for stochastic-dynamic models by Kall and Wallace (1994). Stochasticity is given as requests are uncertain when decisions are taken and subject to a stochastic distribution. It is dynamic as decisions are taken in multiple points in time. Therefore, we draw on an MDP as described in Sect. 4.1 to model the IRP_{BSS}. In the following sections, we formally define the MDP. To this end, we recall the notation of the MDP and its elements as summarized in Table 5.2.

Decision Points

Whenever a vehicle arrives at a station, a decision point $k \in K$ occurs.[2] In every decision point, the provider take a decision concerning the inventory of the current station and the routing of the current vehicle subject to the decision state.

---

[1] We like to note that no user will ever request to rent or to return a bike at the depot.

[2] For the initial decision point $k = 0$, we assume the vehicles to "arrive" at the depot.

Table 5.2 Notation of the Markov decision process

Symbol	Description
Markov decision process	
$K = (0, \ldots, k_{\max})$	Sequence of decision points
$S = \{s_0, \ldots, s_{\max}\}$	Set of decision states
$X_s = \{x_1, \ldots, x_{\max}\}$	Sets of feasible decisions
$X = \bigcup_{s \in S} X_s$	Super set over all decision sets
$s_k^x = (s_k, x),\ \forall s \in S, x \in X_s$	Post-decision states
$\Omega = \{\omega_1, \ldots, \omega_{\max} \mid \omega\colon S \times X \to S\}$	Transition functions
$p_\omega\colon S \times X \to \mathbb{N}_0$	Penalty function subject to ω
State parameters	
$s_k = \left(N, c_n, V, c_v, t_k, f_k^n, f_k^v, n_k^v, a_k^v\right)$	Decision state
$T = (1, \ldots, t_{\max})$	Time horizon
$t_k \in T$	State s_k's point in time
$f_k^n = \left(f_k^{n_0}, \ldots, f_k^{n_{\max}}\right) \in \mathbb{N}_0^{\lvert N \rvert}$	Stations' fill levels in t_k
$f_k^v = \left(f_k^{v_1}, \ldots, f_k^{v_{\max}}\right) \in \mathbb{N}_0^{\lvert V \rvert}$	Vehicles' loads in t_k
$n_k^v = \left(n_k^{v_1}, \ldots, n_k^{v_{\max}}\right) \in N^{\lvert V \rvert}$	Vehicles' current or next stations in t_k
$a_k^v = \left(a_k^{v_1}, \ldots, a_k^{v_{\max}}\right) \in T^{\lvert V \rvert}$	Vehicles' arrival times at n_k^v
Optimization	
$\Pi = \{\pi_0, \ldots, \pi_{\max} \mid \pi: S \to X\}$	Set of policies
$x = (\iota^x, n^x) \in X$	Decision
$\iota^x \in \mathbb{Z}$	Inventory decision
$n^x \in N$	Routing decision

Decision States

Every decision point k is associated with decision state $s_k \in S$. It covers information on the BSS the provider can draw on to make decisions. That means, in first place, attributes addressing the invariable infrastructure are given by the set of stations N, the stations' capacities c_n, the set of vehicles V, and the vehicles' capacities c_v. In second place, s_k also provides variable attributes. The sequence T depicts the time horizon the BSS is managed. Then, $t_k \in T$ is the point in time when the decision point k occurs. For every station n_i, the current fill level is depicted by $f_k^{n_i} \in \mathbb{N}_0$. Analogously, $f_k^{v_i}$ represents the amount of bikes loaded by every vehicle v_i. Vehicle v_i is currently located at or on its way to station $n_k^{v_i} \in N$. It arrives in time $a_k^{v_i}$.

Decisions

Let v_i be the vehicle arriving at station $n_k^{v_i}$ and inducing decision point k. Then, the decisions to be taken concern the inventory of $n_k^{v_i}$ and the routing of v_i. Therefore, every decision $x = (\iota^x, n^x)$ covers two components. The inventory decision $\iota^x \in \mathbb{Z}$ depicts the amount of relocated bikes from the station's view. If the provider decides for the vehicle to pick up bikes, $\iota^x < 0$. If the vehicle delivers bikes, $\iota^x > 0$. The routing decision $n^x \in N$ depicts the station where the vehicle is sent next.

Post-decision States

In the post-decision state s_k^x, the decision $x = (\iota^x, n^x)$ made in s_k has been realized. Therefore, the current station's fill level changes from $f_k^{n_i}$ to $f_k^{n_i} + \iota^x$ and the vehicle's load becomes $f_k^{v_i} - \iota^x$. As time needs to be spent for both relocations and traveling, the vehicle will arrive at n^x in time $t_k + |\iota^x| \cdot \tau_r + \tau_{n_k^{v_i}, n^x}$.

Transition

The next decision point $k + 1$ occurs when the next vehicle arrives at a station. Therefore, time t_{k+1} is the minimum arrival time over all vehicles:

$$t_{k+1} = \min_{v \in V} a_k^v. \tag{5.1}$$

Realizations of requests associated with a specific MDP are depicted by transition functions $\omega \in \Omega$. A transition function ω reveals requests in the time span $(t_k, t_{k+1}]$, i.e., between decision points k and $k + 1$, and alters the stations' fill levels due to successful requests.

Objective

The IRP$_{\text{BSS}}$ can be solved by any policy $\pi \in \Pi$. For any decision state, π yields a decision $\pi(s_k) = x \in X_{s_k}$. When a decision is made, the penalty function p_ω reveals the amount of failed requests between two decision points k and $k + 1$ subject to the decision state, the decision made, and the requests: $p_\omega(s_k, x) \in \mathbb{N}_0$.

The decisions are made with respect to the minimization of failed requests until the end of the time horizon. The final decision point $k_{\max}$ is reached if and only if sending the current vehicle v_i to the depot n_0 is the only feasible decision and every other vehicle is already located at the depot:

$$k = k_{\max} \quad \Leftrightarrow \quad X_{s_k} = \{(0, n_0)\} \ \wedge \ n_k^v = n_0, \ \forall v \in V \setminus \{v_i\}. \tag{5.2}$$

Then, the objective (5.3) is to identify an optimal policy $\pi^* \in \Pi$ minimizing the expected amount of failed request over all decision points conditioned on the initial decision state s_0:

$$\pi^* = \underset{\pi \in \Pi}{\arg\min} \mathbb{E}\left[\sum_{k=0}^{k_{\max}} p_\omega\big(s_k, \pi(s_k)\big) \Big| s_0\right]. \tag{5.3}$$

5.4 Example

After the notation of the IRP$_{\text{BSS}}$ and the MDP have been defined, we provide an example in this section. In the following, we describe the details of the decision state s_k, the decision x, post-decision state s_k^x, the transition ω and the associated penalty, and the next decision state s_{k+1} shown in Fig. 5.1.

The exemplary BSS comprises a set of four stations: $N = \{n_1, n_2, n_3, n_4\}$. For the sake of simplicity, we omit the depot. Stations n_1 and n_2 have capacities of $c_{n_1}, c_{n_2} = 5$. Stations n_3 and n_4 have capacities of $c_{n_3} = 3$ or $c_{n_4} = 4$, respectively. Two vehicles are applied to balance the BSS: $V = \{v_1, v_2\}$. Both vehicles have a capacity of $c_{v_1}, c_{v_2} = 4$.

In the example, vehicle v_1 has just arrived at station n_1 and, therefore, decision point k in time $t_k = 12$ occurs. The associated decision state is depicted on the left-hand side of Fig. 5.1. For the stations, dark boxes stand for available bikes and light boxes for free bike racks. Analogously, for the vehicles, dark boxes represent loaded bikes and light boxes indicate unused capacity. Therefore, the stations' fill levels are $f_k^{n_1} = 4$, $f_k^{n_2} = 0$, $f_k^{n_3} = 2$, and $f_k^{n_4} = 4$. The vehicles both have $f_k^{v_1}, f_k^{v_2} = 1$ bike loaded. Vehicle v_1 induced the decision point due to the arrival at $n_k^{v_1} = n_1$. Therefore, the arrival time corresponds with the decision state's time: $a_k^{v_1}, t_k = 12$. Vehicle v_2 currently travels to $n_k^{v_2} = n_4$ and will arrive in time $a_k^{v_2} = 21$.

As v_1 induces the decision point, we have to make a decision $x = (\iota^x, n^x)$ addressing the inventory of n_1 and the routing of v_1. With respect to the station's capacity

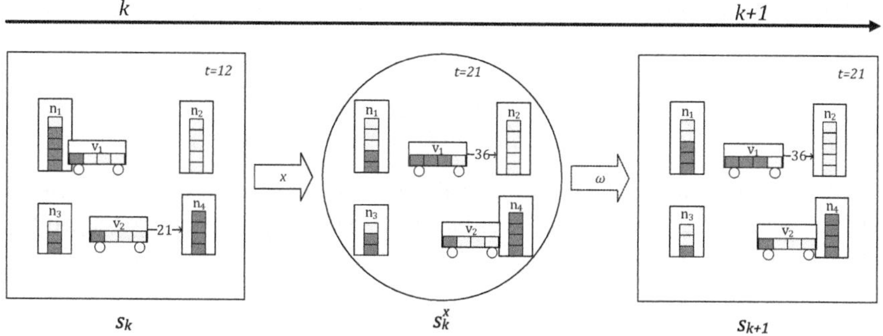

Fig. 5.1 An exemplary MDP of the IPR$_{\text{BSS}}$ (adapted, Brinkmann et al. 2019b)

and fill level, and the vehicle's capacity and load, feasible inventory decisions are picking up one, two, or three bikes, delivering one bike, or resign relocations. Therefore, $\iota^x \in \{-3, -2, -1, 0, 1\}$. Afterwards, the vehicle may remain at its station or travel to any other station: $n^x \in N$. As we are free to combine any realization of ι^x and n^x, the number of feasible decisions $x \in X_{s_k}$ corresponds the cardinality of X_{s_k}: $|X_{s_k}| = 5 \cdot 4 = 20$. In the example, we decide to pick up two bikes ($\iota^x = -2$) and send the vehicle to $n^x = n_2$. The resulting decision is $x = (-2, n_2)$.

The decision state and decision made lead to the post-decision state s_k^x as depicted in the center of Fig. 5.1. Therefore, the fill level of station n_1 is altered to $f_k^{n_1} + \iota^x = 4 + (-2) = 2$ and the load of vehicle v_1 becomes $f_k^{v_1} - \iota^x = 1 - (-2) = 3$. Let $\tau_{n_1,n^x} = 20$ be the travel time to the next station and let $\tau_r = 2$ be the time one relocation requires. Then, v_1 arrives at v_2 in time $t_k + |\iota^x| \cdot \tau_r + \tau_{n_1,n^x} = 12 + |-2| \cdot 2 + 20 = 36$. However, the next decision point s_{k+1} will occur in time $t_{k+1} = \min_{v \in V} a_k^v = 21$ and is induced due to the arrival of v_2 at n_4.

Beforehand, requests between $t_k = 12$ and $t_{k+1} = 21$ are revealed. In the example, at n_2, one rental request fails due to the unavailability of bikes. Although no free bike racks are available at n_4, no return request fails as no user requests to return a bike. Therefore, $p_\omega(s_k, x) = 1$. One return request at n_1 and one rental request at n_3 are successful.

The successful requests are reflected by the next decision state $\omega(s_k, x) = s_{k+1}$ as depicted at the right-hand side of Fig. 5.1. Therefore, the successful return increases the fill level of n_1 by one and the successful rental decreases the fill level of n_3 by one. Summarizing the decisions taken and the successful request, the new fill levels read as follows: $f_{k+1}^{n_1} = 4 + (-2) + 1 = 3$, $f_{k+1}^{n_2} = 0$, $f_{k+1}^{n_3} = 2 - 1 = 1$, and $f_{k+1}^{n_4} = 4$. Now, decisions need to be taken addressing the inventory of station n_4 and the routing of vehicle v_2.

5.5 Challenges

Solving the IRP$_{\text{BSS}}$ to optimality requires to solve the Bellman Equation (5.4) in every decision point k.

$$\pi^*(s_k) = \underset{x \in X_{s_k}}{\arg\min} \mathbb{E}\left[\sum_{k'=k}^{k_{max}} p_\omega(s_{k'}, x) \middle| s_k\right] \tag{5.4}$$

In Sect. 4.1, we point out the resulting challenges for generalized dynamic decision making. Here, we go into details with the state space, the decision space, and the action space of the IRP$_{\text{BSS}}$.

The State Space

The number of different states $s \in S$ is given by $|S|$ and can be depicted by the combination of state attributes. Let us exclusively draw on one specific BSS and, therefore, omit the invariable attributes. The number of points in time is denoted by $t_{\max} + 1$. The combination of stations' fill levels and the vehicles' loads is only limited by the number of bikes within the BSS.[3] The vehicles can be located at any station $n \in N$ and can arrive at any point in time t. Then, $|S| \leq (t_{\max} + 1) \cdot (c_n + 1)^{|N|} \cdot (c_v + 1)^{|V|} \cdot |N|^{|V|} \cdot (t_{\max} + 1)^{|V|}$.

The Solution Space

As discussed in Sect. 4.1, the solution space corresponds to the number of policies where every policy is a sequence of decisions. Therefore, we first quantify the numbers of decisions for every decisions state and second derive the number of policies. The feasible decisions $x \in X_{s_k}$ for a decision state $s_k \in S$ are determined as follows. Let v_i be the current vehicle and $n_k^{v_i}$ be its current station in decision point k. Then, the number of inventory decisions is subject to the current vehicle's capacity c_{v_i} and load $f_k^{v_i}$. We can deliver at most $f_k^{v_i}$ bikes and pick up at most $c_{v_i} - f_k^{v_i}$ bikes. If retaining the current fill level $f_k^{n_k^{v_i}}$ is allowed, we can realize at most $f_k^{v_i} + c_{v_i} - f_k^{v_i} + 1 = c_{v_i} + 1$ different fill levels at the current station $n_k^{v_i}$. The routing decision is about the vehicle's next station. We can send the vehicle to any station $n \in N$ or keep it at the current station. Therefore, $|X_{s_k}| \leq (c_{v_i} + 1) \cdot |N|$. However, each policy does not only have make decisions for a single decision state but for *any* decision state $s \in S$. Therefore, the number of feasible policies is $|\Pi| \leq \left((c_{v_i} + 1) \cdot |N|\right)^{|S|}$.

The Transition Space

The possible realizations of requests $\omega \in \Omega$ are unbounded as any request may occur any time and at any station. Further, requests may occur in any combination.

We experience exponential expansion rates in the state space and in the solution space—the transition space is not penetrable anyway. Further, we can reduce the IRP_{BSS} to the TSP. Given the case that one vehicle with unlimited capacity is applied to conduct relocations. Let the subset of stations in need of relocations to be known. Additionally, the points in time of the actual relocations are irrelevant. Then, we are only restricted by the time horizon. Identifying a tour including all necessary

---

[3]For instance, no BSS has enough bikes such that every station's fill level is at capacity at the same time.

stations without violating the time horizon may result in determining an optimal TSP tour. Determining such a tour requires non-polynomial time (Rosenkrantz et al. 1977). Then, solving the IRP_{BSS} to optimality also requires non-polynomial time. Consequentially, solving the IRP_{BSS} to optimality is most likely not applicable for real-time decision making. Therefore, in Chaps. 6 and 7, we draw on methods of ADP.

Chapter 6
Lookahead Policies

In this chapter, we introduce the lookahead policies originally presented by Brinkmann et al. (2019a, b) to solve the IRP_{BSS}. The LAs base on the blue print introduced in Sect. 4.2.2 but are strongly adapted to the requirements of the IRP_{BSS}.

In Sect. 6.1, we provide an outline, i.e., a verbal description of the idea of the LAs. The formal definitions are given in Sect. 6.2. In Sect. 6.3, the introduced procedures are presented by means of algorithmic formulations.

6.1 Outline

A policy is applied when a decision point occurs in the MDP. Then, the applied policy draws on the associated decisions state's parameters and an algorithmic procedure determining a decision. The policy yields the decision to the MDP. This scheme is sketched in Fig. 6.1. As depicted, an LA comprises two main components: The simulation and the optimization.

In the simulation component, the policy investigates different potential fill levels the current vehicle can realize at its station. The potential fill levels result in different expected amounts of failed requests. In the first place, realizing a certain fill level impacts the corresponding station. However, neighboring stations are also affected since users approach alternative stations if their request fails on first attempt (Rudloff and Lackner 2014). Therefore, in the simulation component, the amounts of failed requests for every station in the course of the lookahead horizon are approximated. An LA draws on *one* of the following two simulation procedures to approximate the failed requests. The procedures differ by the realization of simulations.

Online Simulations

In the first procedure, potential realizations of requests are resampled and the resulting MDP is simulated. In the simulations, the developments of fill levels and failed requests can be observed. We use the average failed requests over all simulations as approximations. Simulating the MDP explicitly includes the users' behaviour and

J. Brinkmann, *Active Balancing of Bike Sharing Systems*, Lecture Notes in Mobility,
https://doi.org/10.1007/978-3-030-35012-3_6

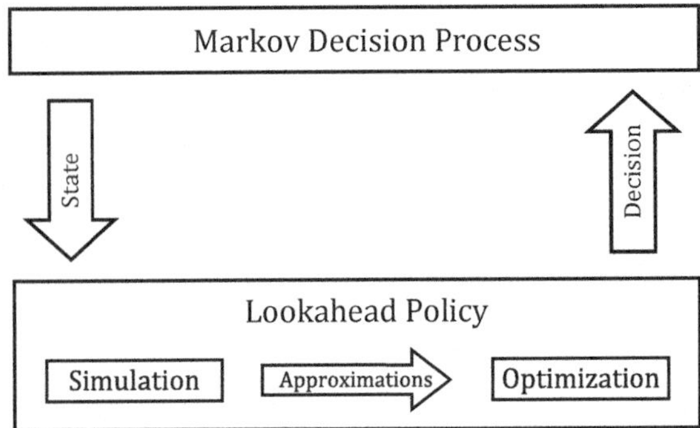

Fig. 6.1 Overview of a lookahead policy (adapted, Brinkmann et al. 2019a)

also the stations' interactions. Further, the samples depict a certain spread. This spread accounts for the deviation of the actual requests of different days. Unfortunately, simulating the MDP requires much computational effort but yields accurate approximations.

Offline Simulations

The second procedure draws on average requests from historical data. Starting from the realized fill levels to investigate at the current station and from the current fill levels at the other stations, we successively update the fill levels with respect to the given average requests to observe the fill levels' developments. If a fill level becomes void or reaches the maximum capacity of the station, the excessive requests count towards the failed requests. Here, the station interactions are neglected. As we draw on average request, also a spread of requests is not considered. The offline simulation component is significantly cheaper in terms of computational effort but yields less accurate approximations.

When the approximation of failed requests is completed, the approximations are forwarded to the optimization component. In the optimization component, the approximations enable anticipation. Here, the inventory and the routing decisions are selected sequentially. First, the inventory decision is determined. The policy selects the decision realizing the fill level that leads to the approximated minimum of failed requests at the current vehicle's station. Second, the routing decision is selected. To this end, the policy assigns every vehicle a station. However, in a decision point, we only have to make a decision for the current vehicle. For the sake of coordination, we make a preliminary assignment for every other vehicle. Therefore, the policy determines the amount of failed request that a vehicles can save at the different stations. The failed requests a vehicle can save at a certain station is subject to the vehicle's capacity, load, travel time from its current station to the station in focus. In certain situations, sending the current vehicle to the station where *it* can save the most requests might be disadvantageous if then another vehicle is sent to an alternative station where only a small amount of failed requests can be saved (e.g., due to long travel time). A coordinated assignment maximizes the saved requests over all vehicles and stations by means of solving an assignment problem.

We limit the observation of fill levels and approximation of failed requests to a predefined lookahead horizon. Ghiani et al. (2009), Voccia et al. (2017) apply lookahead policies for related stochastic-dynamic VRPs and observe low solution qualities if the lookahead horizon is not selected well. If the horizon is too short, only a small amount of requests occurs. Then, the decisions' impacts cannot be distinguished. If the horizon is too long, a huge amount of requests occurs. Then, the decisions' impacts fade out in stochasticity. In Sect. 6.2.2, we provide examples obtained from real-world data to demonstrate that too short as well as too long horizons distort the decision making process. Suitable horizons depend on the exogenous request pattern. In BSSs, the request pattern is subject to spatio-temporal variations. Therefore, a procedure adapting to the request pattern and determining dynamic lookahead horizons is introduced in Chap. 7.

6.2 Definition

In this section, we define the LAs. In Sect. 6.2.1, we define the simulation component. The optimization component is defined in Sect. 6.2.2.

6.2.1 Simulation

Let v be the current vehicle in decision point k and let n_k^v be its station. The simulation investigates three potential fill levels at n_k^v. The fill levels correspond to the percentages $\mu_1 = 25\%$ (*low*), $\mu_2 = 50\%$ (*medium*), and $\mu_3 = 75\%$ (*high*) of the station's capacity $c_{n_k^v}$. The resulting target fill levels are $\mu_1 \cdot c_{n_k^v}$, $\mu_2 \cdot c_{n_k^v}$, and $\mu_3 \cdot c_{n_k^v}$. It might be that not every target fill level is feasible due to the vehicle's capacity c_v and load f_k^v. Therefore, we alter the fill levels accordingly and determine the actual inventory decisions ι_1, ι_2, and ι_3 as shown in Eq. (6.1).

$$\iota_i = \begin{cases} \min\left\{\lfloor \mu_i \cdot c_{n_k^v} \rfloor - f_k^{n_k^v}, f_k^v\right\} & \text{, if } \mu_i \cdot c_{n_k^v} > f_k^{n_k^v} \\ \max\left\{\lfloor \mu_i \cdot c_{n_k^v} \rfloor - f_k^{n_k^v}, f_k^v - c_v\right\} & \text{, if } \mu_i \cdot c_{n_k^v} < f_k^{n_k^v} \\ 0 & \text{, else.} \end{cases} \tag{6.1}$$

The first case occurs if the station's current fill level $f_k^{n_k^v}$ is lower than the target fill level $\mu_i \cdot c_{n_k^v}$. To realize the target fill level, $\mu_i \cdot c_{n_k^v} - f_k^{n_k^v}$ bikes need to be delivered. However, the vehicle can deliver at most f_k^v bikes. Therefore, the inventory decision is set to the minimum of both terms. The second case occurs if the station's current fill level is higher than the target fill level. To realize the target fill level, $f_k^{n_k^v} - \mu_i \cdot c_{n_k^v}$ bikes need to be picked up. The vehicle can pick up at most $c_v - f_k^v$ bikes. In the model, pickups are denoted by $\iota < 0$. Therefore, the terms in the second case of Eq. (6.1) comprise the corresponding negative magnitudes. Then, the inventory

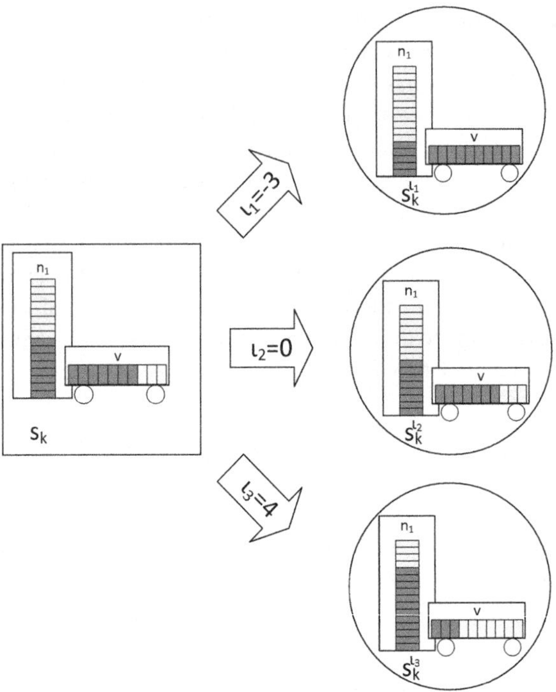

Fig. 6.2 Three inventory decisions and the resulting fill levels

decision is the maximum of both terms. The third case occurs if the station's fill level is equal to the target fill level and no additional relocations are required.

We provide a numerical example of a simulation. As depicted on the left-hand side of Fig. 6.2, v is the current vehicle in decision point k with a capacity $c^v = 10$ and a load of $f_k^v = 7$. Its station is n_k^v with capacity $c^{n_k^v} = 16$ and fill level $f_k^{n_k^v} = 8$. The three target fill levels are $25\% \cdot 16 = 4, 50\% \cdot 16 = 8$, and $75\% \cdot 16 = 12$. Then, the LA determines the actual inventory decisions with respect to Eq. (6.1). The first target fill level cannot be realized as not enough bikes can be picked up by the vehicle. Therefore, the associated inventory decision is $\iota_1 = \max\{4 - 8, 7 - 10\} = -3$, i.e., three bikes will be picked up. The second target fill level is already achieved. Therefore, the second inventory decision is $\iota_2 = 0$ and no bikes will be relocated. The third inventory decision is $\iota_3 = \min\{12 - 8, 7\} = 4$. The resulting fill levels are depicted on the right-hand side of Fig. 6.2.

The resulting failed requests of every inventory decision are approximated either by the online simulation component (Sect. 6.2.1.1) or by the offline simulation component (Sect. 6.2.1.2). When an inventory decision ι is applied, the resulting failed rental and return requests at station n are depicted by $\gamma_{\iota,n}^-, \gamma_{\iota,n}^+ \in \mathbb{R}_0^+$. Failed rentals and returns need to be approximated separately since we later determine how many rentals and returns the individual vehicles can save. We determine the failed requests for the current vehicle's station to make an inventory decision and for every other station to make a routing decision.

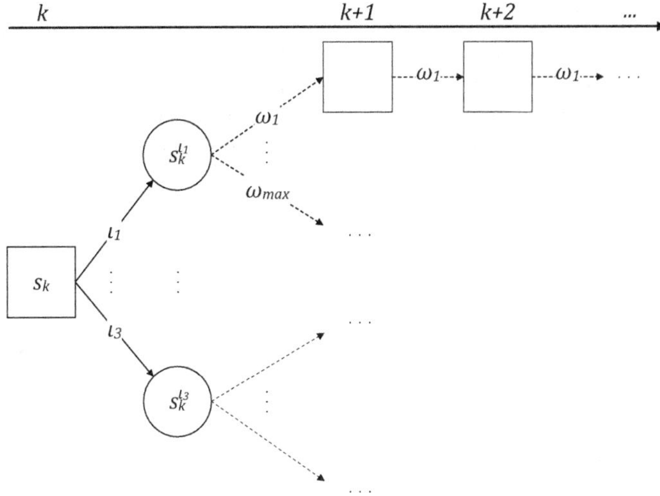

Fig. 6.3 An online lookahead's simulations

6.2.1.1 Online

LAs using the online simulation component (Brinkmann et al. 2019a) are denoted by LA_{on}. An LA_{on} uses stochastic information by resampling realizations of requests and simulates the three inventory decisions at runtime. The procedure of the online simulation is depicted by Fig. 6.3. Starting on the left-hand side in decision point k's state s_k, the online simulation investigates the inventory decisions. For every potential inventory decision ι_i, the resulting post-decision state $s_k^{\iota_i}$ is created.[1] Then, realizations of requests are resampled and the corresponding set of transition functions $\Omega' \subset \Omega$ is generated. For every transition function $\omega_j \in \Omega'$, we simulate the MDP's progress. Unlike rollout algorithms, LAs do not apply any additional decisions in the simulations.[2] Therefore, we omit the post-decision states in Fig. 6.3. Every simulation covers a limited lookahead horizon of length $\delta \in \mathbb{N}$ or stops beforehand if the time horizon is overstepped. Let k^j be a decision point in the simulation of transition function ω_j. Then, the simulation stops when the decision point $k^j_{\max}$ with $s_{k^j_{\max}}$ and $t_{k^j_{\max}} \geq \min\{t_k + \delta, t_{\max}\}$ occurs.

When the simulations are completed, the average failed requests per station with respect to the corresponding inventory decision are determined. Let $p_j^-, p_j^+ : S \times$

[1]We are aware that the notation of post-decision states usually reads s_k^x. As LAs investigate the resulting failed requests with respect to potential inventory decisions and ignores the routing decision in the first step, we use the notation s_k^ι.

[2]In Sect. 8.5.5, we apply rollout algorithms to solve the MDP of the IRP_{BSS}. We conclude, that rollout algorithms are not advantageous compared to the LAs.

$N \rightarrow \mathbb{N}_0$ be the penalty functions associated with ω_j.[3] The failed rental requests are depicted by p_j^- and the failed return requests by p_j^+. Let $\sigma \in \mathbb{N}$ be the number of simulations for every inventory decision. In Eqs. (6.2) and (6.3), the approximated failed rentals $\gamma_{\iota,n}^-$ and returns $\gamma_{\iota,n}^+$ at station n are determined as the average failed rentals and returns over all simulations of the respective inventory decision.

$$\gamma_{\iota,n}^- = \frac{1}{\sigma} \cdot \sum_{j=1}^{\sigma} \sum_{k^j=k}^{k_{\max}^j} p_j^-(s_{k^j}, n) \tag{6.2}$$

$$\gamma_{\iota,n}^+ = \frac{1}{\sigma} \cdot \sum_{j=1}^{\sigma} \sum_{k^j=k}^{k_{\max}^j} p_j^+(s_{k^j}, n) \tag{6.3}$$

6.2.1.2 Offline

LAs using the offline simulation (Brinkmann et al. 2019b) component are denoted by LA_{off}. An LA_{off} draws on average amounts of requests to successively update the stations fill levels. Let $\lambda_t^n \in \mathbb{R}$ be the average amount of requests at station n in time t predetermined and based on historical data.[4] If the majority of requests are rental requests, $\lambda < 0$. If the majority of requests are returns, $\lambda > 0$. If rentals and returns counterbalance each other, $\lambda = 0$. Then, LA_{off} updates the fill levels. Consider a fill level f_t^n at station n in time t.[5] The fill level f_{t+1}^n in time $t + 1$ is approximated according to Eq. (6.4). The term on the right-hand side consists of an inner max function and an outer min function. The max function ensures that the updated fill level does not become negative and the min function ensures that the updated fill level does not exceed the station's capacity.

$$f_t^n = \min\left(\max(f_{t-1}^n + \lambda_t^n, 0), c^n\right) \tag{6.4}$$

Then, the resulting failed rental and return requests over the lookahead horizon of δ are approximated. Equation (6.5) approximates the failed rentals in the lookahead horizon δ at station n given inventory decision ι is realized.[6]

---

[3]Please note: Unlike the penalty functions introduced in Sects. 4.1 and 5.3, the penalty functions in the online simulation refer to states and stations.

[4]In the case studies (Chap. 8), one time unit t corresponds to one minute.

[5]The notation of fill levels in decision point k reads f_k^n. As an LA_{off} investigate the fill level development independently from the MDP, we use the notation f_t^n instead.

[6]Here, an inventory decision ι affects n_k^v only as the offline simulation component neglects station interactions. This is reflected by $\gamma_{\iota_i,n}^- = \gamma_{\iota_j,n}^-$ and $\gamma_{\iota_i,n}^+ = \gamma_{\iota_j,n}^+$, $\forall n \in N \setminus \{n_k^v\}$ and for any two given inventory decisions ι_i, ι_j.

$$\gamma_\iota^-(n) = \sum_{t=t_k+1}^{t_k+\delta} \left|\min(f_{t-1}^n + \lambda_t^n, 0)\right| \tag{6.5}$$

The term in the sum depicts the failed rentals between any time t and the previous point in time. If rentals fail, $f_{t-1}^n + \lambda_t^n$ is negative and is selected by the min function. Then, the absolute magnitudes within the lookahead horizon are summed up. Equation (6.6) approximates the failed returns. The term in the sum depicts the failed returns at station n between any time t and the previous point in time. If rentals fail, $f_{t-1}^n + \lambda_t^n - c^n$ is positive and is selected by the max function.

$$\gamma_\iota^+(n) = \sum_{t=t_k+1}^{t_k+\delta} \max(f_{t-1}^n + \lambda_t^n - c^n, 0) \tag{6.6}$$

When the approximation of failed rentals and returns is completed, the approximations are forwarded to the optimization component as described in the next section. There, we provide developments of fill levels and failed requests in exemplary simulations and show how decisions are made on this basis.

6.2.2 *Optimization*

In the optimization component, an LA uses the approximations of failed requests and, therefore, anticipation is enabled. First, the inventory decision (Sect. 6.2.2.1) is determined. Second, the routing decision (Sect. 6.2.2.2) is selected.

6.2.2.1 Inventory Decision

To make the idea of decision making understandable, we recall the example of Sect. 6.2.1 and show the associated developments in a simulation. These developments are the basis for the LA's decisions.

We evaluate three potential inventory decisions $\iota_1 = -3\,(25\%)$, $\iota_2 = 0\,(50\%)$, and $\iota_3 = 4$ (75%). Figure 6.4 depicts the average developments of fill levels and failed requests in an exemplary offline simulation. Let 14:30 h be the point in time of the decision state. The abscissa depicts the lookahead horizon over the following five hours. The fill levels and failed requests are shown on the ordinate. The solid and dashed lines depict the fill level developments if the target fill levels of 75 and 25% are realized. We omit the fill level development of 50% to keep the figure simple. The dotted line depicts the failed requests resulting from a fill level of 25%. As described, the fill levels of 75 and 25% start with twelve and five bikes. In Fig. 6.4, we see parallel developments of fill levels. Until 16:00 h, the fill levels remain more or less on their initial level as either no requests occur or rentals and returns are equal.

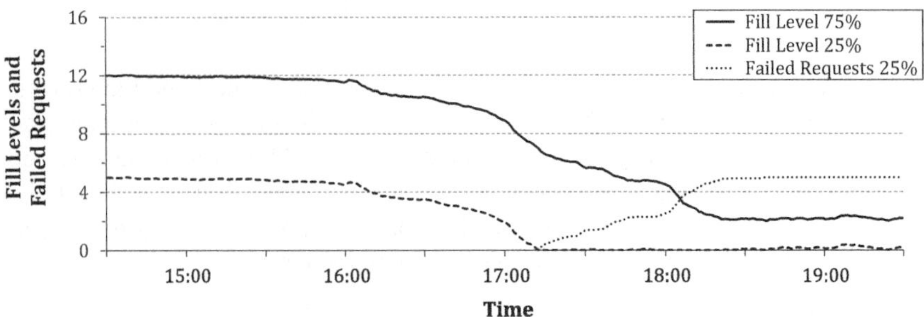

Fig. 6.4 Observed fill levels and failed requests in an exemplary simulation

Then, rental requests take place until 18:30 h. Therefore, the fill levels decrease. At around 17:15 h, users requested to rent approximately five bikes. Therefore, at this point in time, the station runs out of bikes if a fill level of 25% has been realized. From that time on, the approximated amount of failed requests increases. In the end of the lookahead horizon, five rental requests failed in total. Therefore, $\gamma^-_{\iota_1,n^v_k} = 5$. The fill level that started from 75% remains sufficient until the end of the lookahead horizon: $\gamma^-_{\iota_3,n^v_k} = 0$. The fill level of 50% starts with eight bikes and results in two failed rentals, indicated by $\gamma^-_{\iota_3,n^v_k} = 2$. In the simulation, no return requests fail reflected by $\gamma^+_{\iota_1,n^v_k}, \gamma^+_{\iota_2,n^v_k}, \gamma^+_{\iota_3,n^v_k} = 0$.

We select the inventory decision ι^x with respect to the approximated failed requests such that the minimum of requests will fail:

$$\iota^x = \underset{\iota \in \{\iota_1,\iota_2,\iota_3\}}{\arg\min} \left\{\gamma^-_{\iota,n^v_k} + \gamma^+_{\iota,n^v_k}\right\}. \tag{6.7}$$

In the example, $\iota^x = \iota_3$. According to the simulation, ι_3 avoids any failed requests at station n^v_k in the next five hours.

We further observe that with a lookahead horizon of less than 150 min, the LA would not be able to differentiate the inventory decisions since not requests failed for any inventory decision. After 150 and 210 min, the fill levels that started from 25 to 50% result in failed rentals. Therefore, a lookahead horizon longer than 210 min does not provide more information than required. Therefore, the length of the lookahead horizon significantly impacts the solution quality of an LA.

6.2.2.2 Routing Decision

In this section, we describe how routing decisions are taken. As for the inventory decision, the vehicles' capacities and loads are considered. But in essence, we make the decision with respect to the travel times and with respect to the whole vehicle

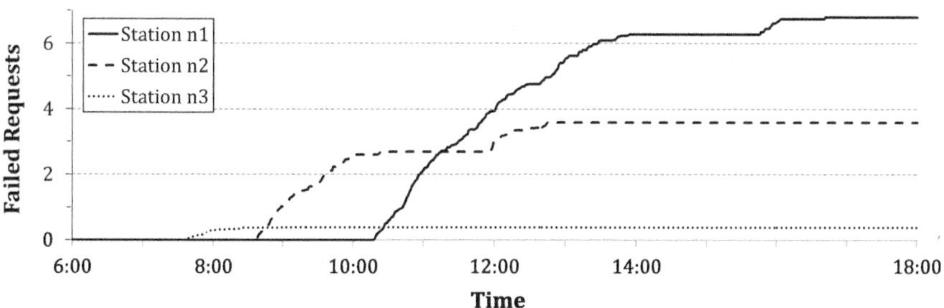

Fig. 6.5 Observed rental requests in an exemplary simulation

fleet. To illustrate the challenges, we provide a number of examples. The challenges reveal that the future decisions of the other vehicles need to be anticipated. Therefore, we assign every vehicle to a station. The resulting assignment problem is solved by a greedy but efficient approach. The assignments are preliminary in the way that we only implement the assignment of the current vehicle. By this, anticipation of the other vehicles' decisions is allowed and flexibility to counteract unexpected requests is preserved.

Challenges

As soon as the inventory decision ι^x is determined, the LA makes the routing decision. To this end, the focus is on the approximated failed rentals and returns resulting from ι^x: $\gamma^-_{\iota^x,n}, \gamma^+_{\iota^x,n}, \forall n \in N$. Also here, the lookahead horizons has an impact on the LA's performance. More precisely, unsuitable lookahead horizons may distort the urgency of relocations at the different stations. Therefore, we discuss varying lookahead horizons applied to the following example. In Fig. 6.5, we depict the failed request of three stations observed in an exemplary lookahead. The lookahead extends from 6:00 over twelve hours to 18:00 depicted on the abscissa. On the ordinate, the failed requests are shown. The lines show the cumulated failed requests observed at stations n_1, n_2, and n_3. In the course of the horizon, requests fail at all three stations. In the end of the lookahead horizon, the approximated amount of failed request is about 6.8 at n_1 as depicted by the solid line. The dashed line depicts that at n_2 the approximated amount of failed requests is 3.6. At n_3, 0.4 requests will fail according to the approximation and shown by the dotted line. Therefore, at n_1 relocations are most urgent if the horizon is applied as depicted. However, failed requests occur at n_2 and n_3 much earlier than at n_1. Therefore, we propose a lookahead horizon to extend at most until 11:00. Until that time, n_2 approximately experiences the most failed requests. Given such a horizon, the LA concludes that relocations are most urgent at n_2. When sending the current vehicle to n_2, there is still enough time to visit n_1 afterwards. Limiting the horizon to 8:00, only n_3 experiences failed requests. Given such a short horizon, n_3 would be classified to be most urgent although only a small amount of requests fails. As a consequence, the requests at n_2 probably cannot be saved if the vehicle is sent to n_3. Therefore, the horizon should at least cover 9:00.

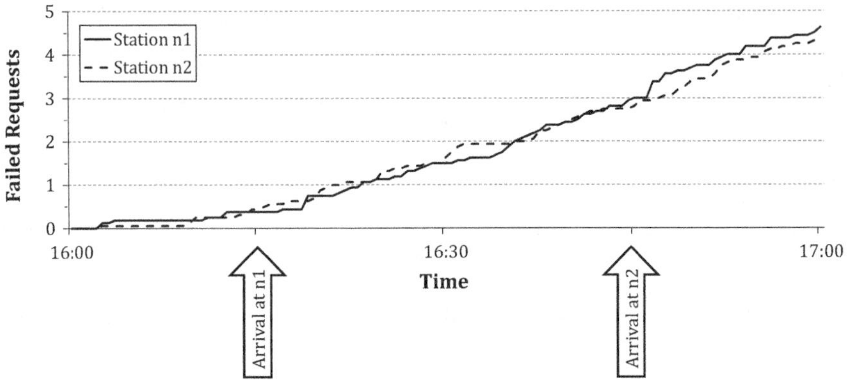

Fig. 6.6 Failed requests in an exemplary simulation (Brinkmann et al. 2019a)

Then, the small amount of failed requests at n_3 is accepted and the large amount of failed requests at n_2 is saved.

As in the previous section, we need to consider the capacity and load of the vehicle v to determine the amount of potentially saved requests. Further, before the vehicle can apply relocations at some station n, it first has to travel from its current station n_k^v to n. This travel time needs to be considered as demonstrated by the following example. Figure 6.6 shows the development of approximated failed requests of station n_1 and n_2. On the abscissa, the exemplary lookahead horizon of one hour between 16:00 and 17:00 is depicted. The ordinate depicts the amount of failed rental requests. The solid and dashed lines represent the developments for n_1 or n_2, respectively. The vehicle has a sufficient number of bikes loaded and, therefore, could save rental requests. The LA now has to select one of these stations to be visited next. At first glance, the developments of failed rentals are more or less the same and, in terms of the objective, it does not matter which station is visited. However, the vehicle is located in neighborhood of station n_1 whereas n_2 is in some other district. The resulting travel times are $\tau_{n_k^v,n_1} = 15$ and $\tau_{n_k^v,n_2} = 45$ min. The associated arrival times are 16:15 and 16:45. We can see that when the vehicle arrives at station n_1, only a small amount of failed requests took place. The majority can be saved by delivering bikes. When the vehicle arrives at n_2, approximately 1.5 rental requests already failed. Therefore, sending the vehicle to n_1 is beneficial.

Certainly, regarding one station, we have to consider every vehicle to determine the potential of saved requests. Figure 6.7 exemplarily demonstrates different arrival times of the vehicles v_1, v_2, and v_3 at station n and resulting impacts. The abscissa depicts the lookahead horizon. The fill level and amount of failed requests are shown on the ordinate. The solid and dashed line indicate the fill levels or approximated failed requests, respectively. The decision point in the example occurs at 14:30. The lookahead horizon has a length of five hours and, therefore, lasts until 19:30. As we can see, the fill level decreases due to rental requests. Just before 16:00, the station

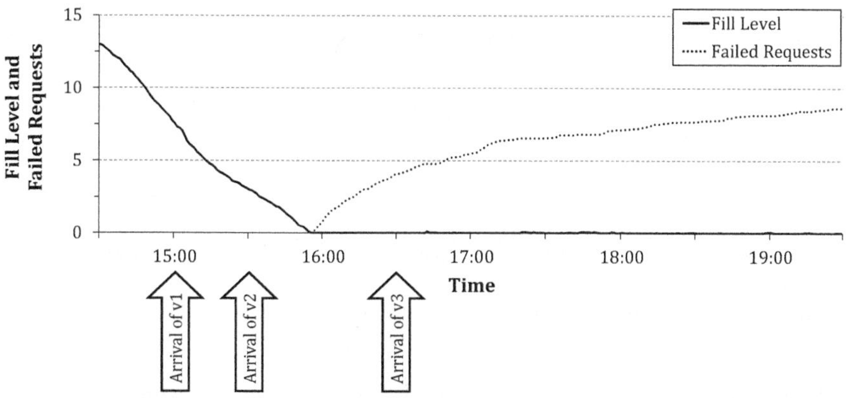

Fig. 6.7 Observed fill level and failed requests in an exemplary simulation (adapted, Brinkmann et al. 2019b)

is empty. From that time on, rental requests fail.[7] Therefore, relocations are required to save the requests. We assume that every vehicle has a sufficient number of bikes loaded. However, the arrival times are different. If vehicle v_1 will be sent to station n, it would arrive at 15:00. Vehicle v_2 would arrive at 15:30 and v_3 at 16:00. Although, v_1 and v_2 have got different arrival times, both would arrive early enough to save all rentals. Vehicle v_3 would arrive when already approximately four rental requests have failed. Therefore, sending v_3 is an unfavorable alternative. From the station's view, we are indifferent whether sending v_1 or v_2.

The two challenges described above reveal the requirement of coordinated dispatching of vehicles. We realize coordinated dispatching by means of solving an assignment problem. In an assignment problem, we assign vehicles to station in order to maximize the saved requests over all vehicles and stations. To this end, we determine the amounts of requests $\gamma_{v,n} \in \mathbb{R}_0^+, \forall v \in V, n \in N$ the vehicles can save at the station by applying Eq. (6.8).

$$\gamma_{v,n} = \max\Bigg\{\underbrace{\min_{[a_k^v+\tau_{n_k^v,n},t_k+\delta]}\left\{\gamma_{\iota^x,n}^-, f_k^v\right\}}_{\text{saved rentals}}, \underbrace{\min_{[a_k^v+\tau_{n_k^v,n},t_k+\delta]}\left\{\gamma_{\iota^x,n}^+, c^v - f_k^v\right\}}_{\text{saved returns}}\Bigg\} \tag{6.8}$$

The amount of saved requests is the maximum of saved rentals and saved returns. These magnitudes can be determined independently. The saved rentals at station n are determined as follows: At first, we consider the failed rentals $\gamma_{\iota^x,n}^-$ approximated by the simulation and given inventory decision ι^x is applied. The maximum of rentals

[7]Notably, we also observe some rare returns indicated by the small increments of the solid line. The corresponding bikes are instantly rented and, therefore, reduce the amount of failed requests.

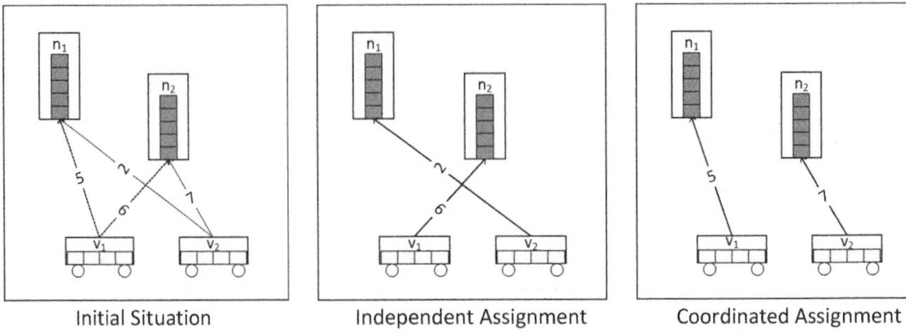

Fig. 6.8 An exemplary assignment problem (adapted, Brinkmann et al. 2019b)

that can be saved is the load f_k^v of vehicle v.[8] Here, we consider the time a_k^v at which v induces its next decision point and the travel time $\tau_{n_k^v,n}$ to n. Therefore, the earliest point in time v can reach n is $a_k^v + \tau_{n_k^v,n}$. The simulation has been limited to $t_k + \delta$. We limit the amount of saved rentals to the time interval $[a_k^v + \tau_{n_k^v,n}, t_k + \delta]$ and, in this way, account for travel times. Then, we consider the failed returns $\gamma_{\iota^x,n}^+$. The maximum of returns that can be saved is the different of capacity c^v and load f_k^v by vehicle v.[9] According to the description above, we limit the amount of saved returns to the time interval $[a_k^v + \tau_{n_k^v,n}, t_k + \delta]$.

In an additional example, we demonstrate the interactions of assignments of vehicles to stations. In the current decision point, we have to make the routing decision for vehicle v_1. The alternatives are stations n_1 and n_2. We also know that the subsequent decision point will be induced by vehicles v_2. On the left-hand side of Fig. 6.8, we show the current decision state. Both stations' fill levels are already at capacity. According to the simulation, return requests will take place at both stations in the near future. The vehicle are empty and, therefore, could save return requests. However, the travel times and the times the failed returns occur are relevant as demonstrated beforehand. Therefore, we focus on the amounts of potentially saved requests. On the left-hand side of Fig. 6.8, the potentially saved requests of v_1 are depicted on the dashed lines. It would save five returns at n_1 and six at n_2. On the dotted lines, the potentially saved requests of v_2 are shown. The vehicle can save two returns at n_1 and seven at n_2. In the center of Fig. 6.8, we show an independent dispatching. As we are only deciding about v_1 in the current decision point, sending it to n_2 in order to save six returns would be the decision made. Then, in the next decision state, the only remaining routing decision is to send v_2 to n_1 in order to save two returns. In total, the independent dispatching results in eight saved returns. Therefore, we propose to realize coordinated dispatching by means of assignments for all vehicles. In a coordinated dispatching, as depicts on the right-hand side of Fig. 6.8, v_1 is assigned to n_1 although it could save one more return at n_2. Then, v_2 is assigned to n_2 with seven saved returns. This assignment results in twelve saved requests and, therefore,

[8]If v is the current vehicle, we use the term $f_k^v - \iota^x$ instead to include the inventory decision.

[9]If v is the current vehicle, we use the term $c^v - f_k^v - \iota^x$ instead to include the inventory decision.

is advantageously. Then, we implement the assignment of v_1 only as it is the vehicle that induced the decision state.

Integer Linear Program

In this section, we formally define the assignment problem as an integer linear program (ILP). In Eq. (6.9), we define decision variables $y_{ij} \in \{0, 1\}, \forall v_i \in V, n_j \in N$ to indicate the assignments of vehicles to stations.

$$y_{ij} = \begin{cases} 1, \text{ if vehicles } v_i \text{ is assigned to station } n_j \\ 0, \text{ else} \end{cases} \tag{6.9}$$

Basing on a general assignment problem formulation by Bazaar et al. (2010), the ILP we use reads as follows:

$$\max \sum_{i=1}^{|V|} \sum_{j=1}^{|N|} y_{ij} \cdot \gamma_{v_i, n_j} \tag{6.10}$$

subject to

$$\sum_{j=1}^{|N|} y_{ij} = 1, \quad \forall v_i \in V \tag{6.11}$$

$$\sum_{i=1}^{|V|} y_{ij} \leq 1, \quad \forall n_j \in N. \tag{6.12}$$

We maximize the amount of saved requests over all vehicles and stations in the objective function (6.10). Constraint (6.11) ensures that every vehicles is assigned to exactly one station. Constraint (6.12) ensures that at most one vehicle is assigned to a specific station.

The ILP can be solved to optimality within polynomial time (Bazaar et al. 2010). However, in the stochastic-dynamic environment, the term "optimal" refers to the unlikely instance when the amounts of failed request occur exactly as approximated. Powell et al. (2000) observe that *deterministic-static* decision making is often counterproductive in *stochastic-dynamic* optimization problems. Therefore, we define a greedy but efficient approach to solve the ILP heuristically.[10]

Matrix Maximum Approach

To solve the assignment problem's ILP, we draw on a matrix maximum approach based on an heuristic approach for the generalized assignment problem by Mattfeld and Vahrenkamp (2014). The idea of the approach is to iteratively identify the vehicle/station assignment that saves the most requests. In every iteration, we ignore

[10]In Sect. 8.7.1, we investigate the impact of the heuristic and the optimal assignment in the stochastic-dynamic environment and confirm the observations of Powell et al. (2000).

the vehicles and stations already assigned. The procedure terminates as soon as the current vehicle is associated with the vehicle/station assignment that saves the most requests.

We realize the approach by creating a matrix $\Gamma \in \mathbb{R}^{|V|\times|N|}$ with elements γ_{v_i,n_j}, $\forall v_i \in V, n_j \in N$:

$$\Gamma = \begin{pmatrix} \gamma_{v_1,n_1} & \cdots & \gamma_{v_{\max},n_1} \\ \vdots & \ddots & \vdots \\ \gamma_{v_1,n_{\max}} & \cdots & \gamma_{v_{\max},n_{\max}} \end{pmatrix}. \tag{6.13}$$

The elements in row i stand for the requests vehicle v_i saves if it is assigned to one of the respective stations. The elements in column j stand for the requests saved at station n_j if one of the respective vehicles is assigned. Then, the approach iteratively searches the maximum element in the matrix. Let v be the vehicle in current decision point and let $\max_\Gamma = \gamma_{v_i,n_j}$ be the maximum entry over all rows and columns in Γ. If the maximum element γ_{v_i,n_j} does not refer to the current vehicle, i.e., $v \neq v_i$, we preliminarily assign v_i to n_j ($y_{ij} = 1$). To avoid conflicting assignments in upcoming iterations, we remove row i and column j from the matrix. Then, v_i and n_j cannot be assigned once more. The approach terminates if the maximum element γ_{v_i,n_j} refers to the current vehicle: $v = v_i$. Then, the routing decision the LA makes is $n^x = n_j$.

Finally, $x = (\iota^x, n^x)$ is the decision taken by an LA.

6.3 Algorithms

In this section, we provide algorithmic formulations for the procedures introduced in the previous sections. In Sect. 6.3.1, we present the algorithm of the LAs. The online and offline simulation components are considered in Sects. 6.3.2 and 6.3.3. In Sect. 6.3.4, the matrix maximum algorithm is examined.

6.3.1 Lookahead Policy

An LA is realized by Algorithm 1. As input, the algorithm receives a decision state that includes all necessary state attributes (Table 5.2, p. 45). The output is a decision the current vehicle realizes afterwards. In lines 1–12, the three potential inventory decisions at the current station are evaluated. The inventory decision is initialized in line 3. If the station's fill level deviates from the target fill level, the inventory decision is defined with respect to the vehicle's load and capacity in 4–10. Then, in line 11, the actual inventory decision is evaluated either by the online or the offline simulation component (Algorithms 2 and 3). When all potential inventory decisions are evaluated, the best performing one is selected in line 13. In line 14, the routing decision is selected by applying the matrix maximum algorithm (4). Finally, the decision is returned in line 15.

Algorithm 1: Lookahead Policy

Input: $s \in S$
Output: $(\iota^x, n^x) \in X$

for all $\mu_i \in \{\mu_1, \mu_2, \mu_3\}$ // For all target fill levels
do
  $\iota_i \leftarrow 0$ // Initialize inventory decision
  if $\mu_i \cdot c_{n_k^v} > f_k^{n_k^v}$ // If fill level lower than target
  then
    $\iota_i \leftarrow \min\{\lfloor \mu_i \cdot c_{n_k^v} \rfloor - f_k^{n_k^v}, f_k^v\}$ // Inventory decision
  else if $\mu_i \cdot c_{n_k^v} < f_k^{n_k^v}$ // If fill level higher than target
  then
    $\iota_i \leftarrow \max\{\lfloor \mu_i \cdot c(n_k^v) \rfloor - f_k^{n_k^v}, c_v - f_k^v\}$ // Inventory decision
  end
  $(\gamma_{\iota_i}^-, \gamma_{\iota_i}^+) \leftarrow \text{Simulation}(s, \iota_i)$ // Apply simulation
end
$\iota^x \leftarrow \arg\min_{\iota \in \{\iota_1, \iota_2, \iota_3\}} \gamma_\iota^-(n_k^v) + \gamma_\iota^+(n_k^v)$ // Select inventory decision
$n^x \leftarrow \text{MatrixMaximumAlgorithm}(\gamma_{\iota^x}^-, \gamma_{\iota^x}^+)$ // Select next station
return (ι^x, n^x)

Algorithm 2: Online Simulations

Input: (s, ι)
Output: $(\gamma_\iota^-, \gamma_\iota^+)$

for all $n \in N$ // Initialization
do
  $\gamma_\iota^-(n) \leftarrow 0$
  $\gamma_\iota^+(n) \leftarrow 0$
end
for all $j \in \{1, \dots, \sigma\}$ // For every simulation
do
  $s_{kj} \leftarrow \omega_j(s, (\iota, n_k^v))$ // Apply inventory decision
  while $k^j \neq k_{max}^j$ // Until end of lookahead horizon
  do
    for all $n \in N$ // For all stations, update failed requests
    do
      $\gamma_\iota^-(n) \leftarrow \gamma_\iota^-(n) + \frac{1}{\sigma} \cdot p_j^-(s_{kj}, n)$
      $\gamma_\iota^+(n) \leftarrow \gamma_\iota^+(n) + \frac{1}{\sigma} \cdot p_j^+(s_{kj}, n)$
    end
    $s_{kj} \leftarrow \omega_j\left(s_{kj}, \left(0, n_k^v\right)\right)$ // Apply no further decisions
  end
end
return $(\gamma_\iota^-, \gamma_\iota^+)$

6.3.2 Online Simulations

The online simulations are realized by Algorithm 2. The input of Algorithm 2 is a decision state and an inventory decision to evaluate. The approximated resulting failed rental and return requests are returned. In lines 1–5, the approximated failed requests are initialized. Then, the actual simulations are conducted in lines 6–18. The first step in every simulation is the application of the given inventory decision in the given decision state in line 8. Then, the simulation's first decision state is obtained from the associated transition function. In lines 9–17, the simulation's decision states are processed until the simulation's final decision state, i.e., until the end of the lookahead horizon. The failed requests resulting from the decision made are captured in lines 11–15. The subsequent decision state is obtained from the transition function in line 16. Technically, the decision made is to neglect further relocations and to stay at the current station. In line 19, the approximated failed requests are returned.

6.3.3 Offline Simulations

Algorithm 3 realizes the offline simulations. As in the algorithm introduced beforehand, the inputs are also a decision state and an inventory decision. The outputs are the approximated failed requests. In lines 1–17, the simulations are separately conducted for every station. At first, the initialization takes place in lines 3–9. We

Algorithm 3: Offline Simulations

Input: (s, ι)
Output: $(\gamma_\iota^-, \gamma_\iota^+)$
1 **for all** $n \in N$ `// For all stations`
2 **do**
3 $\quad \gamma_\iota^-(n) \leftarrow 0$ `// Initialization`
4 $\quad \gamma_\iota^+(n) \leftarrow 0$
5 $\quad t \leftarrow t_k$
6 $\quad f_t^n \leftarrow f_k^n$
7 $\quad$ **if** $n = n_k^v$ **then**
8 $\quad\quad f_t^n \leftarrow f_t^n + \iota$ `// Apply inventory decision`
9 $\quad$ **end**
10 $\quad$ **while** $t \leq t_k + \delta$ `// Until end of lookahead horizon`
11 $\quad$ **do**
12 $\quad\quad t \leftarrow t + 1$ `// Increment point in time`
13 $\quad\quad f_t^n \leftarrow \min(\max(f_{t-1}^n + \lambda_t^n, 0), c^n)$ `// Update fill level`
14 $\quad\quad \gamma_\iota^-(n) \leftarrow \gamma_\iota^-(n) + |\min(f_{t-1}^n + \lambda_t^n, 0)|$ `// Update failed requests`
15 $\quad\quad \gamma_\iota^+(n) \leftarrow \gamma_\iota^+(n) + \max(f_{t-1}^n + \lambda_t^n - c^n, 0)$
16 $\quad$ **end**
17 **end**
18 **return** $(\gamma_\iota^-, \gamma_\iota^+)$

initialize the failed rentals and returns, the point in time of the associated decision state, and the fill level of the associated station. In lines 7–9, the given inventory decision is applied if the associated station is the current vehicle's station. Then, the actual simulations are conducted in lines 10–16. Here, we pass through the lookahead horizon. In every point in time of the horizon, we update the fill level with respect to historical data and the station's capacity. Then, we update the failed rentals in line 14 and the failed returns in line 15. In line 18, the approximations are returned.

6.3.4 *Matrix Maximum Algorithm*

Algorithm 4 realizes the matrix maximum approach. It takes a vehicle and the matrix as input. The matrix' elements depict the expected saved requests for every station and vehicle. The output is the routing decision for the given vehicle, i.e., a station. In lines 1–10, vehicles are iteratively assigned to stations until the given vehicles is assigned. We determine the matrix' maximum element in line 2. Then, we ensure that the station and the vehicle associated with the maximum element are not assigned once more by deleting the elements in the corresponding row and column. Technically, the elements of the corresponding row (lines 3–5) and column (lines 6–8) are set equal to zero. In line 9, we preliminarily assign the station associated with the maximum element to the given vehicle. If the given vehicle is not associated with the maximum element, the procedure is repeated. In line 11, the assignment is returned.

We are aware that the preliminary assignments for all vehicles except for the given are discarded. Only the assigned station for the current vehicle is returned. Therefore, we do not obtain a complete solution for the ILP. However, when the algorithm is called, only the routing decision for the given vehicle is required.

Algorithm 4: Matrix Maximum Algorithm

Input: v, Γ
Output: n^x
1 **do**
2 $\quad \gamma(n_i, v_j) \leftarrow \max_\Gamma \gamma(n, v)$ `// Determine matrix' maximum element`
3 $\quad$ **foreach** $n \in N$ **do**
4 $\quad\quad \gamma(n, v_j) \leftarrow 0$ `// Delete elements in row`
5 $\quad$ **end**
6 $\quad$ **foreach** $v \in V$ **do**
7 $\quad\quad \gamma(n_i, v) \leftarrow 0$ `// Delete elements in column`
8 $\quad$ **end**
9 $\quad n^x \leftarrow n_i$ `// Preliminary assignment`
10 **while** $v \neq v_j$
11 **return** n^x

Chapter 7
Dynamic Lookahead Horizons

As observed by Ghiani et al. (2009), Voccia et al. (2017), the length of a lookahead horizon has a significant impact on the solution quality when solving a stochastic-dynamic optimization problem. Therefore, the lookahead horizon must neither be too short nor too long. More precisely, the lookahead horizon needs to match the request pattern. In BSSs, we are facing spatio-temporal request patterns (Sect. 2.4). Therefore, in this chapter, we approach lookahead horizons for LAs changing in the course of the day. We define a dynamic LA (DLA) to be an LA with lookahead horizons individually selected for certain periods of the time horizon. To this end, we use a procedure originally introduced by Brinkmann et al. (2019a).

In Sect. 7.1, we provide a verbal description on how to determine dynamic lookahead horizons. The associated procedures are formally defined in Sect. 7.2. In Sect. 7.3, we present algorithmic formulations of the procedures.

7.1 Outline

To realize dynamic lookahead horizons, we subdivide the overall time horizon into periods. Then, we determine a suitable lookahead horizon out of a set of candidate horizons for every period. Therefore, a sequence of horizons defines a DLA. In our case studies in Chap. 8, we subdivide the time horizon into 24 periods each of one hour length. We consider seven candidate lookahead horizons. The horizons extend from zero hours, resulting in a myopic policy, to six hours. In the stated setting, the number of distinct DLAs can be calculated as follows: In the first 19 periods—the 19th period corresponds to the time span from 18:00 h to 18:59 h—we can select horizons up to the maximum of six hours. Therefore, seven candidate horizons are feasible. In the 20th period, i.e., between 19:00 h and 19:59 h, the horizon of six hours length would overstep the time horizon by more than one hour. Therefore, the longest feasible horizon in the 20th period has a length of five hours and, therefore, six candidate horizons are feasible. In the subsequent periods, the maximum horizons are accordingly shorter. Therefore, the number of feasible DLAs is $7^{19} \cdot 6! > 8.2 \cdot 10^{18}$.

J. Brinkmann, *Active Balancing of Bike Sharing Systems*, Lecture Notes in Mobility,
https://doi.org/10.1007/978-3-030-35012-3_7

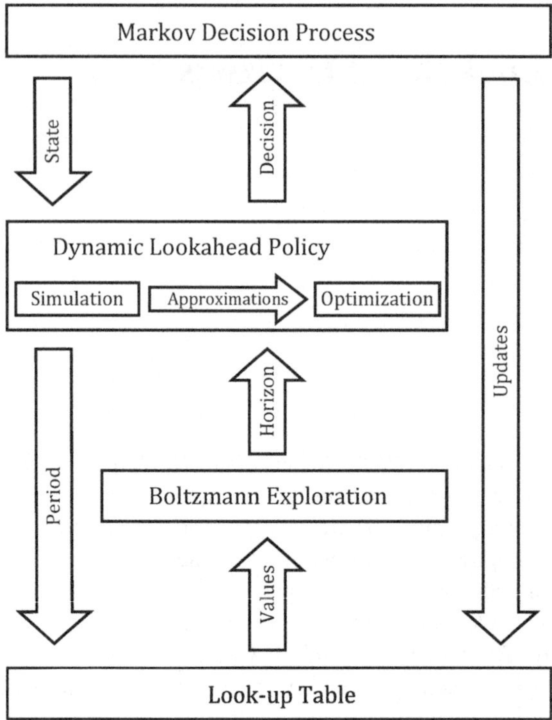

Fig. 7.1 An iteration of a value function approximation's approximation phase in combination with a dynamic lookahead policy (adapted, Brinkmann et al. 2019a)

Due to this large number, identifying the optimal sequence of lookahead horizons analytically is most likely not possible. Therefore, we draw on a procedure that is based on an VFA as introduced in Sect. 4.2.3. It identifies a suitable sequence of horizons by approximating the expected failed requests when applying a certain lookahead horizon in a certain period.

The VFA is subdivided into an approximation phase and an application phase. Figure 7.1 provides an overview on one iteration of the approximation phase of the VFA. In fact, we combine an VFA and an DLA. Therefore, Fig. 7.1 is a combination of Figs. 4.6[1] and 6.1.[2] When a decision point occurs in the current trajectory's MDP, the associated decision state is forwarded to the DLA. The DLA covers a simulation and an optimization component as described in Chap. 6. Before the DLA can carry out simulations, a lookahead horizon needs to be selected. As we draw on horizons for distinct periods, the period associated with the time of the decision state is passed to the look-up table. In the look-up table, the approximated future failed requests for every period/horizon combination are stored as values. The values of the current

[1]"An Iteration of an VFA's Approximation Phase", Sect. 4.2.3, p. 38.

[2]"Overview of a Lookahead Policy", Sect. 6.1, p. 51.

period's candidate horizons are forwarded to the Boltzmann exploration. Unlike the VFA introduced in Sect. 4.2.3, here, we do not approximate the values of decisions but values of parameters for an independent policy—namely an DLA. Another difference is that we do not implement the approximate Bellman Equation selecting always the horizon with the lowest value. When approximating the values, applying pure exploitation often leads to poor solution quality (Powell 2011) especially when the solution space is huge (Rothlauf 2011). Therefore, we use the Boltzmann exploration instead. It controls exploration and exploitation of the DLAs' solution space by occasionally selecting horizons with higher values in the beginning of the approximation phase. The selected horizon is used in the simulation component of the DLA. Then, the resulting approximations are used in the optimization component. The decision taken by the optimization is used in the current decision point of the MDP. When the current trajectory's MDP is completed, the observed failed requests are used to update the values in the look-up table. When the approximation is completed, in every period, the DLA applies the lookahead horizon with the lowest value.

7.2 Definition

In this section, we formally define the procedure to identify suitable sequences of lookahead horizons for DLAs. We define sequences of lookahead horizons and describe the impact on periods formally in Sect. 7.2.1. In Sect. 7.2.2, we define the VFA. The Boltzmann exploration is introduced in Sect. 7.2.3. We define the symbols shown in Table 7.1 one after another and describe their interactions.

Table 7.1 Notation of the value function approximation

Symbol	Description
Sequences of Lookahead Horizons	
$\Delta = \{\delta_0, \ldots, \delta_{\max}\} \subset \mathbb{N}_0$	Set of candidate lookahead horizons
$\mathrm{P} = \{\rho_0, \ldots, \rho_{\max}\}$	Set of periods
Value Function Approximation	
$\nu\colon \mathrm{P} \times \Delta \to \mathbb{R}_0^+$	Value function
$\tilde{\nu}\colon \mathrm{P} \times \Delta \to \mathbb{R}_0^+$	Approximated value function
$\mu\colon \mathrm{P} \times \Delta \to \mathbb{N}_0$	Observed failed requests in a trajectory
$\alpha\colon \mathrm{P} \times \Delta \to \mathbb{N}_0$	Occurrences of hour/horizon pairs
Boltzmann Exploration	
$\eta \in \mathbb{N}_0$	Number of traversed trajectories
$\varepsilon\colon \mathrm{P} \to \mathbb{R}_0^+$	Coefficient to control exploitation and exploration
$\phi\colon \mathrm{P} \times \Delta \to \mathbb{R}^+$	Probabilities of choosing a lookahead horizon

7.2.1 Sequences of Lookahead Horizons

In the first step, we define a set of candidate lookahead horizons $\Delta = \{\delta_0, \ldots, \delta_{\max}\}$ and a set of periods $\mathrm{P} = \{\rho_0, \ldots, \rho_{\max}\}$. The periods are used to partition the time horizon.[3] Technically, the periods are disjoint subsets of the MDP's sequence of decision points K:

$$\rho_i, \rho_j \subset K \ \wedge \ \rho_i \cap \rho_j = \emptyset, \ \forall \rho_i, \rho_j \in \mathrm{P} \ \text{ and } \ \rho_0 \cup \cdots \cup \rho_{\max} = K. \tag{7.1}$$

A decision point k is assigned to a certain period ρ with respect to its point in time t_k. The lookahead horizon used in period ρ_i is denoted as δ^{ρ_i}. Therefore, an DLA is defined by a sequence of lookahead horizons as shown in Eq. (7.2):

$$(\delta^{\rho_0}, \ldots, \delta^{\rho_{\max}}) \in \Delta^{|\mathrm{P}|}. \tag{7.2}$$

Then, the objective (7.3) is to identify an optimal sequence of lookahead horizons $(\delta^{\rho_0}, \ldots, \delta^{\rho_{\max}})^*$ such that the expected failed requests over every period and over every period's decision points conditioned on the initial decision state s_0 are minimized:

$$(\delta^{\rho_0}, \ldots, \delta^{\rho_{\max}})^* = \underset{(\delta^{\rho_0}, \ldots, \delta^{\rho_{\max}}) \in \Delta^{|\mathrm{P}|}}{\arg\min} \ \mathbb{E}\left[\sum_{\rho \in \mathrm{P}} \sum_{k \in \rho} p_\omega\big(s_k, \pi^{\delta^{\rho}}(s_k)\big) \Big| s_0 \right]. \tag{7.3}$$

In period ρ_i, we can depict the resulting failed requests of the selection of horizon δ_j by the value $v(\rho_i, \delta_j)$. Let $s_{k'}$ denote the initial decision state of period ρ_i and let π^{δ_j} be an LA with lookahead horizon δ_j. Then, the values are recursively defined in Eq. (7.4). The left term depicts the expected failed requests in the current period ρ_i conditioned on $s_{k'}$. The right term is necessary since the selection of a certain lookahead horizon in a certain period may pay off in later periods. If the current period is not the last, the right term depicts the expected failed requests over every future period. We assume that in future periods, the horizon resulting in the minimum of failed request is selected.

$$v(\rho_i, \delta_j) = \underbrace{\mathbb{E}\left[\sum_{k \in \rho_i} p\big(s_k, \pi^{\delta_j}(s_k)\big) \Big| s_{k'} \right]}_{\text{current period}} + \underbrace{\begin{cases} \min_{\delta \in \Delta} v(\rho_{i+1}, \delta) & \text{, if } \rho_i \neq \rho_{\max} \\ 0 & \text{, else} \end{cases}}_{\text{future period}} \tag{7.4}$$

[3]Therefore, introducing periods results in state space aggregation. In Sect. 10.2, we discuss advanced state space aggregation techniques.

Assuming that the values are available, the optimal sequence of lookahead horizons can be achieved by selecting lookahead horizon δ^{ρ} with minimal value in every period ρ according to the Bellman Equation (7.5).

$$\delta^{\rho} = \arg\min_{\delta \in \Delta} \nu(\rho, \delta), \ \forall \rho \in \mathrm{P} \tag{7.5}$$

Due to the curses of dimensionality (Sect. 5.5), we cannot determine the values exactly or identify an optimal sequence of lookahead horizons, respectively. Therefore, in the next section, we draw on an VFA to approximate the values.

7.2.2 *Value Function Approximation*

The VFA we use to determine suitable sequences of lookahead horizons is based on the blue print introduced in Sect. 4.2.3. Therefore, we define the approximation phase in Sect. 7.2.2.1 and the application phase in Sect. 7.2.2.2.

7.2.2.1 Approximation Phase

In the approximation phase, we approximate the values $\nu(\rho, \delta)$. To this end, we define an approximated value function $\tilde{\nu}$ such that $\nu(\rho, \delta) \approx \tilde{\nu}(\rho, \delta), \forall \rho \in \mathrm{P}, \delta \in \Delta$. We initialize the approximated values with zero. Then, trajectories of the MDP are traversed. The decisions are taken by an DLA. The DLA's sequence of lookahead horizons $(\delta^{\rho_0}, \ldots, \delta^{\rho_{\max}})$ is selected by the Boltzmann exploration. In a nutshell, in every period, for every lookahead horizon we determine a probability to be selected. Therefore, horizons supposed to perform inferior can be selected in the approximation phase. However, the Boltzmann exploration is worth a dedicated discussion in Sect. 7.2.3

Let the latest trajectory base on ω. Then, the observed failed requests resulting from selecting lookahead horizon δ^{ρ_i} in period ρ_i are depicted by $\mu(\rho_i, \delta^{\rho_i}) \in \mathbb{N}_0$.

$$\mu(\rho_i, \delta^{\rho_i}) = \sum_{k \in \rho_i} p_{\omega}\left(s_k, \pi^{\delta^{\rho_i}}(s_k)\right) + \begin{cases} \mu(\rho_{i+1}, \delta^{\rho_{i+1}}) & , \text{if } \rho_i \neq \rho_{\max} \\ 0 & , \text{else} \end{cases} \tag{7.6}$$

The term on the left-hand side of Equation (7.6) depicts the failed requests in the associated period. The term on the right-hand side refers to future periods. We use the observed failed requests to update the approximated values according to Eq. (7.7).[4] The goal is to use the average failed requests over all trajectories as approximated values. Therefore, we use the old approximation *until* the current trajectory and

[4]Just a note for a clearer understanding: In our setting, we have 24 periods. Therefore, after every trajectory, we update 24 values.

the new observation *in* the current trajectory and calculate the weighted sum. Here, $\alpha_{\rho,\delta} \in \mathbb{N}_0$ depicts the number of selections of horizon δ in period ρ over all trajectories traversed up to the current trajectory.

$$\tilde{v}(\rho_i, \delta^{\rho_i}) := \underbrace{\frac{\alpha_{\rho_i,\delta^{\rho_i}} - 1}{\alpha_{\rho_i,\delta^{\rho_i}}} \cdot \tilde{v}(\rho_i, \delta^{\rho_i})}_{\text{old approximation}} + \underbrace{\frac{1}{\alpha_{\rho_i,\delta^{\rho_i}}} \cdot \mu(\rho_i, \delta^{\rho_i})}_{\text{new observation}} . \tag{7.7}$$

The approximation phase ends as soon as the termination criterion occurs. To this end, we compare the sequence of horizons selected by the Boltzmann exploration with those that would have been selected by the Bellman Equation (7.8). If in 1.000 sequential trajectories the Boltzmann exploration and the Bellman Equation select the same sequence of horizons, exploration is not realized any more. Therefore, we stop the approximation phase.

7.2.2.2 Application Phase

In the application phase, we apply the approximated values $\tilde{v}(\rho, \delta)$ in the approximate Bellman Equation (7.8) to select suitable lookahead horizons:

$$\delta^{\rho} = \arg\min_{\delta \in \Delta} \tilde{v}(\rho, \delta), \;\; \forall \rho \in \mathrm{P}. \tag{7.8}$$

The selected horizons are then used for the sequence $(\delta^{\rho_0}, \ldots, \delta^{\rho_{\max}})$. Eventually, the resulting policy is denoted by DLA.

7.2.3 Boltzmann Exploration

The Boltzmann exploration we use here is based on the one proposed by Powell (2011). Unlike the Bellman Equation drawing on pure exploitation, the Boltzmann exploration controls exploitation and exploration. In the context of dynamic decision making, exploitation means to select the decision with the minimum of immediate and future penalties. In the approximation phase of an VFA, the future penalties, i.e., the values, are approximated. Further, trajectories are traversed by drawing on incomplete approximated values. Thus, in the early stage of the approximation phase, the approximations are inaccurate. For instance, they may be distorted due to improbable transitions in some trajectories. The penalties of every trajectory have an huge impact on the approximations if the number of traversed trajectories is small. Therefore, exploration is realized by selecting decisions by means of a randomized procedure.

In this way, the approximation phase is supported by exploration, i.e., by occasionally selecting inferior decisions.[5]

In the Boltzmann exploration, controlling exploitation and exploration is realized as follows.[6] In a period ρ, every lookahead horizon δ_i is assigned a probability $\phi_{\rho,\delta_i} \in (0, 1)$ to be selected which is defined in Eq. (7.9).

$$\phi_{\rho,\delta_i} = \frac{\exp\left(-\tilde{v}(\rho, \delta_i) \cdot \varepsilon_\rho\right)}{\sum\limits_{\delta \in \Delta} \exp\left(-\tilde{v}(\rho, \delta) \cdot \varepsilon_\rho\right)} \tag{7.9}$$

A certain probability reflects the ratio of the associated approximated value to the sum over all values.[7] However, the values are modified. To increase the values' impacts on the actual probability, the values $\tilde{v}(\rho, \delta)$ are multiplied with a parameter $\varepsilon_\rho \in \mathbb{R}_0^+$ and leveraged by the natural exponential function $\exp\colon \mathbb{R} \to \mathbb{R}^+$. The parameter ε_ρ controls exploitation and exploration.

In essence, if the control parameter is large, the Boltzmann exploration enforces exploitation. Exploration is forced if the control parameter is small. In the beginning of the approximation phase, the approximated values are inaccurate. Then, exploration is desired to guarantee a certain number of trajectories applying the assumable inferior horizons. In the end of the approximation phase, the approximations are already accurate. Then, exploitation is desired as selecting inferior horizons would not make sense. Further, if the approximated values for a certain period have more or less the same magnitudes, exploitation is used in order to verify the assumable superior horizon. When the magnitudes differ strongly, exploration is desired to verify the assumable inferior horizons. We define ε_ρ in period ρ according to this requirements in Eq. (7.10).

$$\varepsilon_\rho = \begin{cases} 0 & \text{, if } \max_{\delta \in \Delta} \tilde{v}(\rho, \delta) = \min_{\delta \in \Delta} \tilde{v}(\rho, \delta) \\ \frac{\eta \cdot 0.01}{\max_{\delta \in \Delta} \tilde{v}(\rho,\delta) - \min_{\delta \in \Delta} \tilde{v}(\rho,\delta)} & \text{, else.} \end{cases} \tag{7.10}$$

The first case occurs when the maximum and the minimum values are equal and, therefore, all values are equal. Then, pure exploration is applied, i.e., the probabilities are the same for all horizons. Usually, this case occurs only in the first trajectory. In the second case, we draw on the number of trajectories traversed so far denoted by $\eta \in \mathbb{N}_0$ and on the difference of maximum and minimum values. The parameter ε_ρ grows with η in the numerator.[8] This results in exploration in the beginning and exploitation in the end of the approximation phase. The denominator comprises

[5]"Inferior" in the sense of higher approximated values.

[6]In deterministic-static optimization, the Boltzmann exploration is well known as a part of the local search algorithm simulated annealing (Aarts and Korst 1989).

[7]As small values point out better horizons, negative magnitudes are used. In maximization problems, one does not need negative magnitudes.

[8]The factor 0.01 is a tuning parameter. Preliminary experiments revealed that good results are achieved with this setting.

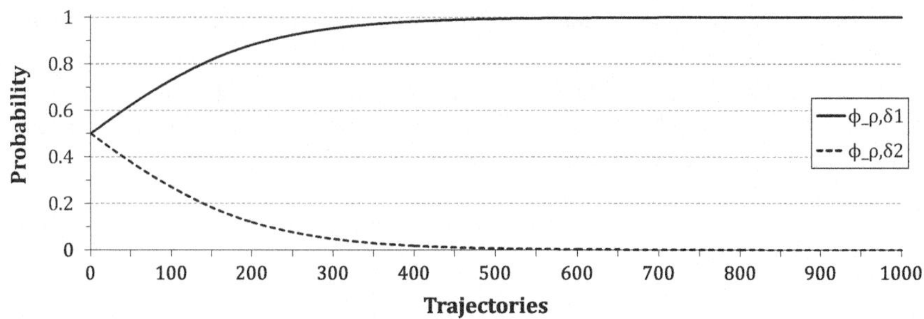

Fig. 7.2 Probabilities of two exemplary values in the course of an approximation phase

the difference of maximum and minimum values. When the difference is large, ε_ρ becomes small and exploration is enforced. When the difference is small, ε_ρ becomes large and exploitation is enforced.

We further describe the concept in two examples. In the first example, we investigate the impact of the number of trajectories on the probabilities. Let $\tilde{\nu}(\rho, \delta_1) = 1$ and $\tilde{\nu}(\rho, \delta_2) = 2$ be the values of two horizons δ_1 and δ_2. The values remain constant and so does their difference. Then, in Fig. 7.2, we observe the associated probabilities ϕ_{ρ,δ_1} and ϕ_{ρ,δ_2} in the course of an approximation phase, i.e., with growing η. On the abscissa, the number of traversed trajectories η are shown. The ordinate indicates the probabilities. ϕ_{ρ,δ_1} is depicted by the solid line and ϕ_{ρ,δ_2} by the dashed line. We can see that in the beginning of the approximation phase, both probabilities are equal indicating exploration. Then, ϕ_{ρ,δ_1} increases and ϕ_{ρ,δ_2} decreases. At some point, ϕ_{ρ,δ_1} is dominating and ϕ_{ρ,δ_2} fades out indicating pure exploitation.

In the second example, we investigate the impact of the difference of maximum and minimum values on the probabilities. Again, $\tilde{\nu}(\rho, \delta_1) = 1$ and $\tilde{\nu}(\rho, \delta_2) = 2$. We consider a third horizon δ with $\tilde{\nu}(\rho, \delta_3) = 1 + 1.1^i$. We start with the exponent $i = 0$ resulting in $\tilde{\nu}(\rho, \delta_2) = \tilde{\nu}(\rho, \delta_3)$. We increase the difference of maximum and minimum values by increasing i stepwise by 0.1 until $i = 100$. Figure 7.3 depicts the resulting probabilities when keeping $\tilde{\nu}(\rho, \delta_1)$ and $\tilde{\nu}(\rho, \delta_2)$ constant and increasing $\tilde{\nu}(\rho, \delta_3)$ by means of increasing i. Therefore, the difference of maximum and minimum value steadily increases where $\tilde{\nu}(\rho, \delta_1)$ is the minimum and $\tilde{\nu}(\rho, \delta_3)$ is the maximum. On the abscissa, the exponent i is shown. Again, the ordinate indicates the probabilities. ϕ_{ρ,δ_1} is depicted by the solid line, ϕ_{ρ,δ_2} by the dashed line, and ϕ_{ρ,δ_3} is indicated by a dotted line. We observe pure exploitation when $i = 0$ since ϕ_{ρ,δ_1} is dominating. Then, when the difference is increased, ϕ_{ρ,δ_1} and ϕ_{ρ,δ_2} draw near. ϕ_{ρ,δ_3} remains insignificant. At some point, we are indifferent between δ_1 and δ_2 which indicates exploration.

In both examples, the approximated values $\tilde{\nu}(\rho, \delta_1)$ and $\tilde{\nu}(\rho, \delta_2)$ are constant and the probabilities develop as desired.

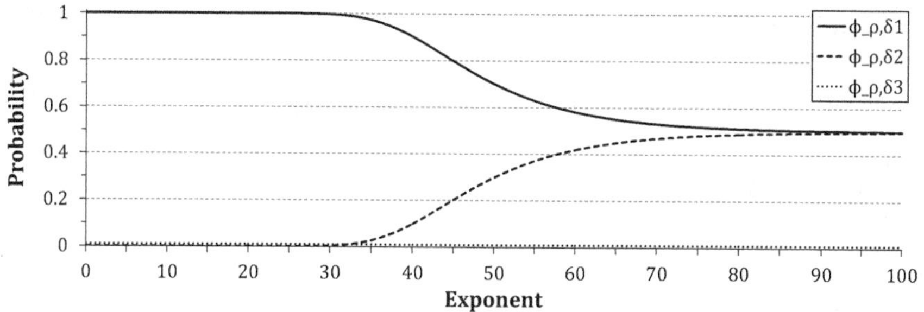

Fig. 7.3 Probabilities of three exemplary values in the course of an approximation phase

7.3 Algorithms

In this section, we present algorithmic formulations of the VFA (Sect. 7.3.1) and the Boltzmann exploration (Sect. 7.3.2).

7.3.1 Value Function Approximation

The VFA is realized in Algorithm 5. The algorithm's inputs are the BSS infrastructure, the set of periods P and the set of considered lookahead horizons Δ. The output is the approximated value function $\tilde{\nu}$. The approximated values, the numbers of selections and the number of traversed trajectories are initialized in lines 1–6. Then, in lines 7–41, the trajectories are traversed until a termination criterion holds. At first, the number of trajectories is increased in line 8. In lines 9–13, the lookahead horizons used in the next trajectory are selected. The actual selection is realized by the Boltzmann exploration (Algorithm 6). For every period $\rho \in$ P, the selected horizon reads δ^{ρ}. Additionally, the associated numbers of selections are incremented. The initial decision point is initialized in line 14. Then, we initialize the stations' fill levels. To this end, first, we set the fill levels equal to zero in lines 15–17. Second, in lines 18–25, the bikes are distributed. The number of bikes corresponds to 50% of the sum of bike racks over all stations. For every bike, a station is selected using the uniform distribution. The fill levels of the selected stations are increased if the selected station's fill level is not at capacity. The vehicles are initialized in lines 26–29. Every vehicle starts at the depot with an empty load. Finally, the initial decision state can be initialized in line 30. We process the decision points of the trajectory's MDP in lines 32–36. To this end, we first determine the current decision state's period.[9] Then, we apply an LA (Algorithm 1) with the selected lookahead horizon to the decision state, apply the transition to obtain the next decision state and go to the next decision point. When the final decision state has been processed, the values

[9]In the case studies (Chap. 8), we use periods of length one hour.

Algorithm 5: Value Function Approximation

Input: $N, c_n, V, c_v, \mathrm{P}, \Delta$
Output: $\tilde{v}$
for all $\rho \in P, \delta \in \Delta$ // Initializations
do
  $\tilde{v}(\rho, \delta) \leftarrow 0$
  $\alpha_{\rho,\delta} \leftarrow 0$
end
$\eta \leftarrow 0$
repeat
  $\eta \leftarrow \eta + 1$ // Count trajectories
  for all $\rho \in P$ // For all hours select lookahead horizon
  do
    $\delta^\rho \leftarrow \mathrm{boltzmann}(\rho, \eta, \tilde{v})$ // Apply Boltzmann exploration
    $\alpha_{\rho,\delta^\rho} \leftarrow \alpha_{\rho,\delta^\rho} + 1$
  end
  $k \leftarrow 0$ // Initial decision point
  for all $n \in N$ **do**
    $f_0^n \leftarrow 0$ // Initialize fill levels
  end
  $b \leftarrow \frac{1}{2} \cdot \sum_{n \in N} c_n$ // Distribute bikes
  while $0 < b$ **do**
    $n \leftarrow \mathrm{random}\{n_1, \ldots, n_{\max}\}$ // Select random station
    if $f_0^n < c_n$ **then**
      $f_0^n \leftarrow f_0^n + 1$ // Increase selected station's fill level
      $b \leftarrow b - 1$
    end
  end
  for all $v \in V$ **do**
    $n_0^v \leftarrow n_0$ // Vehicles start at the depot
    $f_0^v \leftarrow 0$ // Vehicles are initially empty
  end
  $s_0 \leftarrow (N, c_n, V, c_v, 0, f_0^n, f_0^v, n_0^v, a_0^v)$ // Initial decision state
  while $k \neq k_{max}$ // Until final decision point
  do
    $i \leftarrow \lfloor \frac{t_k}{60} \rfloor$ // Determine period
    $s_{k+1} \leftarrow \omega\left(s_k, \pi^{\delta^{\rho_i}}(s_k)\right)$ // Apply DLAs decision and transition
    $k \leftarrow k + 1$ // Go to next decision point
  end
  for all $\rho \in P$ // Update values for all periods
  do
    $\tilde{v}(\rho, \delta^\rho) \leftarrow \frac{\alpha_{\rho,\delta^\rho} - 1}{\alpha_{\rho,\delta^\rho}} \cdot \tilde{v}(\rho, \delta^\rho) + \frac{1}{\alpha_{\rho,\delta^\rho}} \cdot \sum_{k=\bar{k}^\rho}^{k_{\max}} p\left(s_k, \pi^{\delta^\rho}(s_k)\right)$
  end
until *STOP*
return $\tilde{v}$

of all periods and the selected horizons are updated in lines 37–40. When the last trajectory has been traversed, the approximated value function is returned in line 42.

7.3.2 Boltzmann Exploration

Algorithm 6 realizes the Boltzmann exploration. It returns a lookahead horizon δ for a given period ρ. The parameter ε is defined in lines 1–4. Then, for every horizon, the probabilities to be selected are determined in lines 5–8. The actual selection is made by a roulette wheel selection in lines 9–16. To this end, we use the uniform distribution and draw $r \in \mathbb{R}$ from the interval $[0, 1)$. Then, we iteratively sum up the probabilities of horizons δ_i in w. As soon as w exceeds r, the current horizon δ_i is selected. The selected horizon is returned in line 17.

Algorithm 6: Boltzmann Exploration

Input: $\rho, \eta, \tilde{v}$
Output: δ
1 $\varepsilon_\rho \leftarrow 0$ `// Parameter controlling exploitation and exploration`
2 **if** $\max_{\delta \in \Delta} \tilde{v}(\rho, \delta) \neq \min_{\delta \in \Delta} \tilde{v}(\rho, \delta)$ **then**
3 $\quad \varepsilon_\rho \leftarrow \frac{\eta \cdot 0.01}{\max_{\delta \in \Delta} \tilde{v}(\rho,\delta) - \min_{\delta \in \Delta} \tilde{v}(\rho,\delta)}$
4 **end**
5 **for all** $\delta_i \in \Delta$ `// Determining probabilities`
6 **do**
7 $\quad \phi(\rho, \delta_i) \leftarrow \frac{\exp\left(-\tilde{v}(\rho,\delta_i)\cdot\varepsilon_\rho\right)}{\sum_{\delta \in \Delta} \exp\left(-\tilde{v}(\rho,\delta)\cdot\varepsilon_\rho\right)}$
8 **end**
9 $r \leftarrow \text{random}[0, 1)$
10 $i \leftarrow 0$
11 $w \leftarrow \phi(\rho, \delta_i)$
12 **while** $\neg(r \leq w)$ `// Roulette wheel selection`
13 **do**
14 $\quad i \leftarrow i + 1$
15 $\quad w \leftarrow w + \phi(\rho, \delta_i)$
16 **end**
17 **return** δ_i

Chapter 8
Case Studies

In this chapter, we present the case studies to evaluate the policies introduced in the previous chapter. If not stated otherwise, the computations are carried out on an Intel Core i5-3470 with 3.2 GHz and 32 GB RAM. Our approaches are implemented in Java 8.0u121. In Sect. 8.1, we describe the real-world data sets we use. Section 8.2 addresses the actual generation of instances. We describe the implementation of transitions, i.e., the handling of relocations and requests in Sect. 8.3. Benchmark policies are defined in Sect. 8.4. In Sect. 8.5, we present the parametrization of our policies as well as the benchmark policies. The results are presented in Sect. 8.6. In Sect. 8.7, the solutions' structures are analyzed.

8.1 Real-World Data

The instances we are using in the case studies are based on real-world data by the BSSs "Nice Ride" in Minneapolis (MN/USA) and "Bay Area BikeShare" in San Francisco (CA/USA). The provider Motivate International Inc. supplies data sets including detailed information about the stations and about recorded trips. Before the data can be used, data preprocessing steps are necessary as described in Sect. 8.1.1. In Sect. 8.1.2, an overview on the resulting data is presented.

8.1.1 Data Preprocessing

The information about the stations cover identifiers, the capacities as well as the latitudes and the longitudes. The information about recorded trips cover the points in time and the associated stations where the rental and return requests occurred.

J. Brinkmann, *Active Balancing of Bike Sharing Systems*, Lecture Notes in Mobility,
https://doi.org/10.1007/978-3-030-35012-3_8

We implement the data preprocessing procedure proposed by Vogel et al. (2011) to reveal the request pattern. The individual steps of the procedure are described in this section.

The BSS Bay Area comprises three disjoint subsystems East Bay, San Francisco, and San José. Here, we only draw on the subsystem of San Francisco. As the usage pattern in BSSs differs between weekdays and weekends (Sect. 2.4.2), we only draw on trips starting on weekdays. Then, we exclude each trip that lasts shorter than one minute or longer than one hour. Trips with duration shorter than one minute presumably have been recorded when a user observes a failure and returns the bike immediately. In this case, no trip occurred. Trips with a duration longer than one hour are recorded, e.g, when users fail to return their bike. Such trips do not reflect the actual request pattern. We further only consider the months with the most intensive user activity. Due to expansion and maintenance, some stations are added or removed from BSSs. Therefore, we only include stations that experienced at least one rental and one return per month.

8.1.2 *Resulting Data Set*

The key information about the investigated BSSs is summarized in Table 8.1. For Minneapolis, we consider data from the year 2015. For San Francisco, the data has been recorded in 2014. In the considered period, the BSS in Minneapolis has been expanded. We only consider the primal stations. San Francisco comprises a fifth of the number of stations in Minneapolis. In both BSSs, the station capacities are more or less the same. We observe that the sets of recorded trips have significantly been reduced to the sets of considered trips. With respect to the number of weekdays in the respective months, in Minneapolis on average twice as many trips occur per day compared to San Francisco. However, the number of requests per day and station is more than twice as high in San Francisco compared to Minneapolis.

Table 8.1 Characteristics of investigated bike sharing systems

BSS	Minneapolis	San Francisco
Name	Nice ride	Bay area
Provider/Data source	Motivate International Inc. (2015, 2016)	
Year	2015	2014
Considered months	June–September	July–October
Weekdays in considered months	88	89
Considered stations	169/190	35/35
Station capacities	15–35	15–27
Considered trips	197,726/483,229	100,728/292,752
Avg. trips per day	2,246.89	1,131.78
Avg. requests per day and station	26.59	64.67

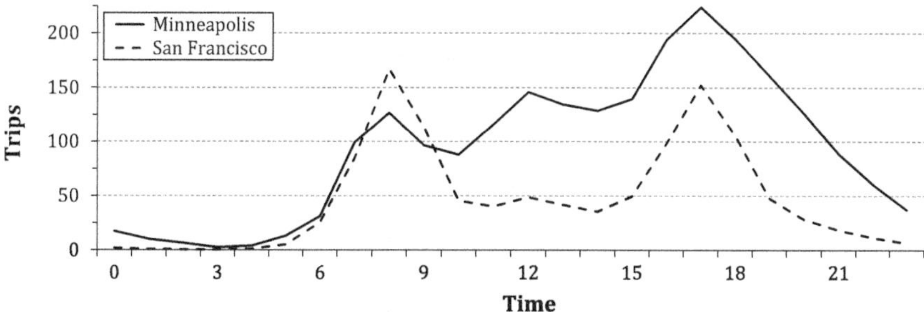

Fig. 8.1 Temporal distributions of trips (adapted, Brinkmann et al. 2019b)

The temporal distributions of trips are shown in Fig. 8.1. On the abscissa, the hours of a day are shown. The average number of trips are depicted on the ordinate. The solid line depicts the average numbers of trips in Minneapolis. The dashed line refers to San Francisco. In both BSSs, peaks occur in the morning, at noon, and in the afternoon. In Minneapolis, the noon and afternoon peaks are higher than the respective previous one. In San Francisco, the morning peak is slightly higher than the one in the afternoon. The noon peak is much lower than the other two. As commuter usage is observed in most BSSs (Sect. 2.4.2), we assume that the morning and afternoon peaks are due to users cycling to their workplaces or back home. In Minneapolis, leisure usage is also significant (O'Brien et al. 2014). Therefore, we assume that at noon and in the afternoon, commuter and leisure activities take place and, as a consequence, the peaks are higher.

8.2 Instances

To generate the actual instances, we resample sets of trips. To this end, we draw trips with replacement from the sets of considered trips. The probabilities to be drawn are equally distributed. By drawing as described, we can create $197{,}726^{2{,}246}$ different instances for Minneapolis, and $100{,}728^{1{,}131}$ for San Francisco. The different instances preserve the respective request patterns. We are also using this procedure for the simulation runs in the LAs with online simulation component. For every simulation run, we resample a distinct set of trips. As we are only interested in the BSS' progress within the lookahead horizon, every trip that does not occur within the lookahead horizon is discarded. For the LAs with offline simulation component, we determine $\lambda_t^n, \forall n \in N, t \in T$ as the average differences of rental and return requests over all weekdays in the set of considered trips and for every minute and station.

We use vehicles with a capacity of 20 bikes. To determine the travel times, we assume the vehicles and cyclists to move with a speed of $15\frac{\text{km}}{\text{h}}$. For vehicles, this is a relatively low speed. However, it accounts for driving as well as parking. The pedestrians' speed is $5\frac{\text{km}}{\text{h}}$. The distance between the stations is euclidean. Every

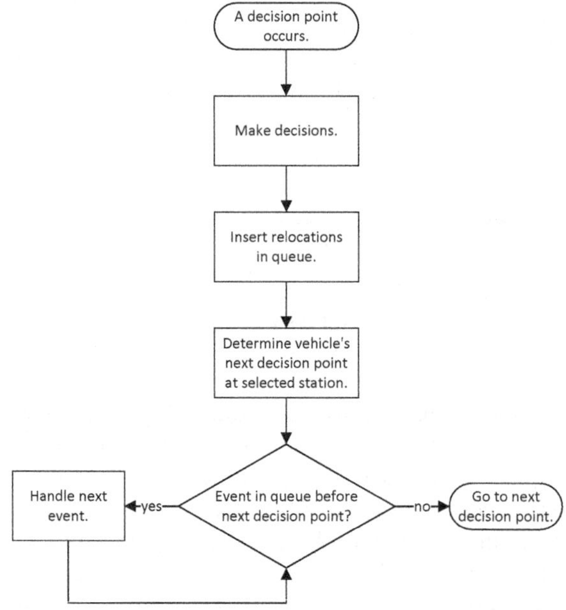

Fig. 8.2 Processes between two decision points

relocated bike requires a service time of two minutes. For the total number of bikes, we draw on 50% of the sum of bike racks over all stations which is a common ratio in practice (O'Brien et al. 2014). That is $50\% \cdot 3.020 = 1{,}510$ in Minneapolis and $\lfloor 50\% \cdot 665 \rfloor = 332$ in San Francisco. The resulting requests per bike and day are $\frac{2{,}246.89 \cdot 2}{1{,}510} = 2.98$ in Minneapolis and $\frac{1{,}131.78 \cdot 2}{332} = 6.82$ in San Francisco.

8.3 Transition

In this section, we provide a description of the transition between two decision points. We realize the transition by implementing an event-based process. That is, we insert events, i.e., relocations and requests, in a queue ordered by the associated points in time. Then, the events are handled sequentially.

Before the MDP begins, we insert the rental requests of every trip in the respective set of trips in the queue. Figure 8.2 depicts the procedures between two decision points. When a decision point occurs, we first make a decision by applying a policy. Then, we insert the relocations in the queue of events according to the inventory decision. To guarantee feasibility, we use one relocation event for every relocated bike.[1] We determine the current vehicle's next decision point with respect to the

[1]For instance, we create two unique relocation events if $\iota = 2$.

inventory and routing decisions. The events in the queue are handled as long as the first event takes place before the next decision point occurs.[2] The procedure is repeated until the final decision point is processed.

Figure 8.3 depicts the procedure of handling events. We first distinguish between relocations and requests. Further, pick up and delivery relocations and rental and return requests are handled differently. If one bike should be picked up, one bike must be available. Analogously, if one bike should be delivered, one bike rack must be free. If the respective requirement is not fulfilled, the event is discarded.[3] If a bike rental is requested, one bike must be available. If a bike is available at the origin station, one bike is rented. That is, the fill level of the station is decreased by one and one return request is created. The return request's point in time depends on the Euclidean distance between the origin and destination station and on the speed. Then, the return request is added to the event queue. If the rental fails as no bike is available at the origin station n_o to cycle to the destination station n_d, the user approaches an alternative origin station n_{o_2}. As shown in Eq. (8.1), n_{o_2} is the station closest to n_o that does not extend the travel time to n_d.

$$n_{o_2} = \underset{n \in N}{\arg\min} \left\{ \tau(n_o, n) \mid \tau(n, n_d) \leq \tau(n_o, n_d) \right\} \tag{8.1}$$

Then, a second rental request is created at n_{o_2} with respect to the travel time between n_o and n_{o_2} and added to the queue. When a bike return is requested, one bike rack must be free. If a bike rack is free at the destination station n_d, the station's fill level is increased by one and the trip ends.
If the return fails, the user approaches the nearest station n_{d_2} alternatively:

$$n_{d_2} = \underset{n \in N}{\arg\min} \left\{ \tau(n_d, n) \right\}. \tag{8.2}$$

Then, a second return request is created at n_{d_2} considering the travel time and added to the queue.

8.4 Benchmarks

We depict the advantages of applying the policies introduced in Chaps. 6 and 7 by comparing their results with those of two benchmark policies from literature. To this end, we define a myopic policy in Sect. 8.4.1 and an anticipating policy in Sect. 8.4.2.

[2]Of course, future requests in the queue are not revealed to the policy applied.

[3]Indeed, that happens rarely.

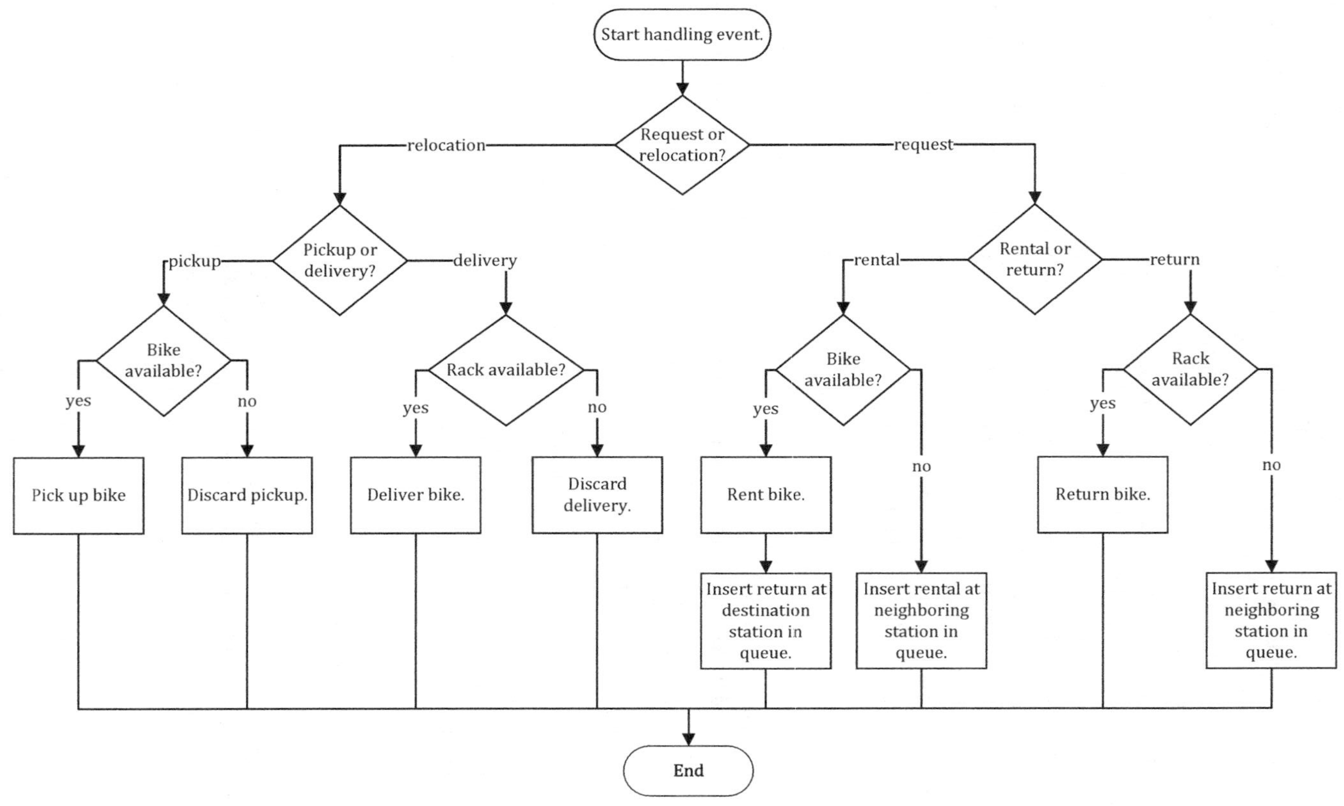

Fig. 8.3 Procedure of handling events (adapted, Brinkmann et al. 2019a)

8.4.1 *Safety Buffer-Tending Relocation Policy*

We apply the safety buffer-tending relocation policy (STR) by Brinkmann et al. (2015) as an exemplary myopic policy. The main idea is to provide safety buffers of bikes and free bike racks at every station to prevent failed requests. The request pattern is ignored. Therefore, STR acts myopically. Similar policies have been applied by Coelho et al. (2014a) for conventional IRPs.

8.4.1.1 Outline

STR draws on safety buffers, i.e., minimum numbers of bikes and free bike racks. If one of these two safety buffers is violated, the associated station is in need of relocations. For the inventory decision, the current vehicle aims on realizing the safety buffers at its current station with minimum effort. If one of the safety buffers is violated, the vehicle picks up or delivers bikes such that the respective safety buffer is just about to be satisfied. Then, the vehicle is sent to the nearest imbalanced station.

8.4.1.2 Definition

The safety buffers are determined by a parameter $\beta \in (0, 1)$. Let c_n be the capacity of station n, then the safety buffers of bikes and free bike racks are $\beta \cdot c_n$. Therefore, in decision point k, the fill level f_k^n is desired to fulfil $\beta \cdot c_n \leq f_k^n$ and $f_k^n \leq (1 - \beta) \cdot c_n$ as well. In other words, $\beta \cdot c_n$ and $(1 - \beta) \cdot c_n$ define a target interval for the fill level.

Decision point k is induced when vehicle v arrives at station n_k^v. Then, STR makes decision $x = (\iota^x, n^x)$ as follows. We determine the inventory decision ι^x as shown in Eq. (8.3). The first case occurs when the fill level $f_k^{n_k^v}$ violates the safety buffer of bikes. Then, the number of bikes to deliver is the difference between lower interval boundary and fill level. The number of deliveries is limited to the number of bikes currently loaded by the vehicle. The second case occurs when the safety buffer of free bike racks is violated. Here, we use negative magnitude as $\iota^x < 0$ depicts pick ups. The number of bikes to pick up is the difference between fill level and upper interval boundary. We limit the pick ups to the number of bikes the vehicles can additionally load. The third case occurs when both safety buffers are satisfied. Then, no bikes will be relocated.

$$\iota^x = \begin{cases} \min\left\{\beta \cdot c_{n_k^v} - f_k^{n_k^v}, f_k^v\right\} & \text{, if } \neg\left(\beta \cdot c_{n_k^v} \leq f_k^{n_k^v}\right) \\ \max\left\{(1-\beta) \cdot c_{n_k^v} - f_k^{n_k^v}, f_k^v - c_v\right\} & \text{, if } \neg\left(f_k^{n_k^v} \leq (1-\beta) \cdot c_{n_k^v}\right) \\ 0 & \text{, else} \end{cases} \tag{8.3}$$

The routing decision n^x is made, by means of assigning every station $n \in N$ a score $\varrho_n \in \mathbb{R}_0^+$. The score reflects the need of relocations and the travel time. For a station n, the score is determined as shown in Eq. (8.4).

$$\varrho_n = \begin{cases} \frac{1}{\tau_{n_k^v,n}} & \text{, if } \neg\left(\beta \cdot c_n \leq f_k^n\right) \wedge 0 < f_k^v - \iota^x \\ \frac{1}{\tau_{n_k^v,n}} & \text{, if } \neg\left(f_k^n \leq (1-\beta) \cdot c_n\right) \wedge f_k^v - \iota^x < c_v \\ 0 & \text{, else} \end{cases} \tag{8.4}$$

If the safety buffer for bikes is violated and the vehicle has at least one bike loaded after the inventory decision ι^x is applied, the score is equal to the inverse travel time between the vehicle's current station n_k^v and n. If the safety buffer for free bike racks is violated and the vehicle can load at least one additional bike after the inventory decision ι^x is applied, the score is again equal to the inverse travel time. If both safety buffers are satisfied, the station is not considered for relocations. Then, the nearest imbalanced station is visited next:

$$n^x = \arg\max_{n \in N \setminus \{n_k^v\}} \{\varrho_n\}. \tag{8.5}$$

If the safety buffers are too small, certain stations may be visited too late to save requests. If the safety buffers are too large, certain stations will be visited although others are in need.

8.4.2 *Rollout Algorithms*

Additionally, we compare the results of the LAs with those of rollout algorithms. In this way, we can demonstrate the advantageous of using an anticipating policy adapted to the application over a general applicable anticipating policy. We implement the blue print by Goodson et al. (2017) introduced in Sect. 4.2.2. Though, we limit the number of decisions to evaluate. Following the idea of LAs, we aim on realizing *high* (75%), *medium* (50%), and *low* (25%) fill levels. With respect to the vehicle's load and capacity, we guarantee feasibility according to Eq. (6.1) in Sect. 6.2.1 (p. 53). If idling at the current station is not allowed, the vehicle may be sent to one of the other 168 stations in Minneapolis and one of the 34 other stations in San Francisco. Therefore, in every decision state, the rollout algorithms evaluate $3 \cdot 168 = 504$ decisions in Minneapolis and $3 \cdot 34 = 102$ in San Francisco. STR as well as DLA_{off} are computationally cheap and, therefore, meet the requirements placed on a base policy. Hence, we apply rollout algorithms with both policies.

8.5 Parametrization

In this section, we identify suitable parameters for the policies. The parameters are the safety buffers for STR (Sect. 8.5.1), the numbers of simulation runs for the LAs using online simulations (Sect. 8.5.2), the lookahead horizons for SLAs (Sect. 8.5.3)

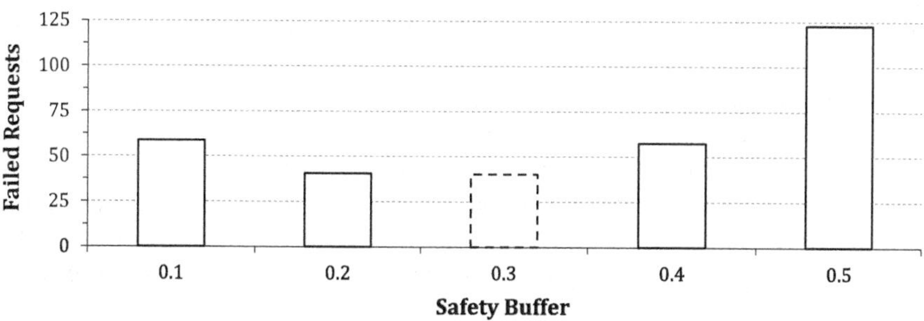

Fig. 8.4 Determining the safety buffers of STR

and DLAs (Sect. 8.5.4), and the base policies and the numbers of simulation runs for the rollout algorithms (Sect. 8.5.5). We identify individual parameters for vehicle fleets of one to four vehicles, for coordinated and independent dispatching, and for Minneapolis and San Francisco. We describe preliminary experiments to identify the parameters before the experiments for actual evaluations are conducted. For every experiment, we investigate 1,000 instances and depict the average results achieved.

8.5.1 Safety Buffer-Tending Relocations

In this section, we identify the safety buffers for STR. We investigate safety buffers $\beta \in \{0.1, \ldots, 0.5\}$. The safety buffers are stepwise increased by 0.1 over four steps. We conduct experiments for one to four vehicles. For two to four vehicles, we apply coordinated and independent dispatching. Therefore, we have $5 \cdot (1 + 3 \cdot 2) = 35$ experimental setups for every BSS.

Figure 8.4 exemplarily depicts the failed requests achieved by STR with various safety buffers and four vehicles in Minneapolis. On the abscissa, the safety buffers are shown. The ordinate depicts the failed requests. We observe that the failed requests are decreased if the safety buffer is increased from 0.1 to 0.3. Increasing the safety buffer from 0.3 to 0.5 comes with increased failed requests. Therefore, in this example, $\beta = 0.3$ is the best safety buffer depicted by the dashed-framed bar.

The best performing safety buffers for every BSS, vehicle fleet size, and for coordinated as well as independent dispatching are depicted in Table A.1 of Appendix A. The results for every experimental setup can be found in Appendix B.

8.5.2 Online Simulations

To identify an appropriate number of simulation runs in the online simulation component of the LAs, we undertake experiments with SLA_{on} and coordinated dispatching.

We use the identified number also for DLA_{on} and for the approximation phase of the VFA. A high number of simulation runs promises a small amount of failed requests on the one hand but requires high runtimes on the other hand. Therefore, the number of simulation runs needs to be chosen well.

We use lookahead horizons of length $\delta \in \{60, \ldots, 360\}$ and $\sigma \in \{1, \ldots, 64\}$ simulation runs. We increase the lookahead horizons five times stepwise by 60. The numbers of simulation horizons are doubled over six stages. One to four vehicles are dispatched for Minneapolis as well as San Francisco. Therefore, we investigate $6 \cdot 7 \cdot 4 = 168$ experimental setups for every BSS.

The resulting failed requests and runtimes are shown in Tables A.2 and A.3 of Appendix A. Here, we are only interested in the horizons achieving the minimum of failed requests for every specific vehicle fleet size, number of simulation runs, and BSS. Therefore, Fig. 8.5 shows the number of simulation runs on the abscissa and the failed requests in the ordinate. The bars depict the average failed requests over the four vehicles fleet sizes achieved by the best performing simulation horizons. The dark bars refer to Minneapolis and the light bars refer to San Francisco. We observe that the failed requests are decreased if the number of simulation runs is increased. However, the more simulation runs are applied, the more runtime is required to solve one instance. In Fig. 8.6, the sums of runtimes to solve an MDP over the vehicles fleet sizes and horizons for the different numbers of simulation runs and for both BSS are depicted. We observe that the required runtime grows with the simulation runs applied.

Consequentially, we have to find a compromise to solve the tradeoff between failed requests and runtime. We observe that for both BSSs, the average failed requests are decreased by less than one when doubling the number of simulation runs from 32 to 64. The sum of runtimes increases from 315.03 to 497.81 s for Minneapolis and from 120.32 to 184.05 s for San Francisco. As a high runtime may limit the approximation phase of the DLAs' VFA, we use $\sigma = 32$ simulation runs for the online simulation component for every LA.

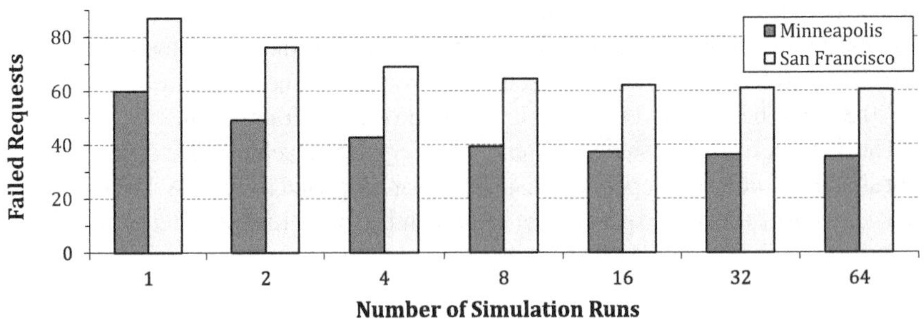

Fig. 8.5 Determining the number of simulation runs of SLA_{on}: failed requests

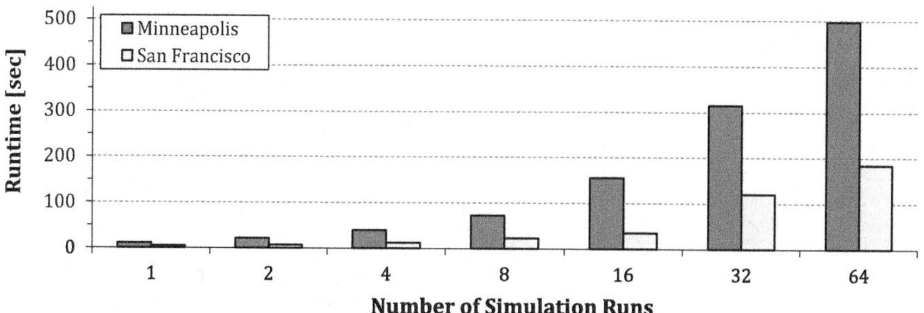

Fig. 8.6 Determining the number of simulation runs of SLA_{on}: runtime

8.5.3 Static Lookahead Policies

We identify suitable lookahead horizons for the SLAs by means of experiments. As aforementioned, both too short and too long horizons lead to many failed requests.

For SLA_{on}, we investigate horizons of length $\delta \in \{60, \ldots, 360\}$ minutes. Generally speaking, SLA_{off} requires a longer horizon. Therefore, we investigate horizons of length $\delta \in \{60, \ldots, 720\}$ minutes. Again, the horizons are stepwise increased by 60 min. We conduct experiments for one to four vehicles, for coordinated and independent dispatching, and for both BSSs. Therefore, we have $(6 + 12) \cdot (1 + 3 \cdot 2) = 126$ experimental setups for every BSS.

We exemplarily discuss the impact of the lookahead horizon for a vehicle fleet size of four in Minneapolis. The general behaviour can be illustrated by this examples. Figures 8.7 and 8.8 depict the failed requests achieved by SLA_{on} and SLA_{off}, respectively. The abscissas show the lookahead horizons, the ordinates show the resulting failed requests. In Fig. 8.7 for SLA_{on}, we observe that the failed requests can be decreased when increasing the horizon from 60 to 180 min. When the horizon is further increased, the failed requests are also increasing. Therefore, the horizon of length $\delta = 180$ min leads to the minimum of failed requests in this example shown by the dashed-framed bar. In Fig. 8.8 for SLA_{off}, the failed requests are decreasing until the horizon is increased to $\delta = 420$ min. The failed requests are increasing if longer horizons are applied.

The best performing static lookahead horizons for every BSS, vehicle fleet size, and for coordinated as well as independent dispatching are depicted in Table A.1 of Appendix A. The results for every experimental setup can be found in Appendix B.

8.5.4 Dynamic Lookahead Policies

We apply the VFA to determine dynamic lookahead horizons as described in Chap. 7. Due to the huge computational requirements of the approximation phase, we here apply an AMD Threadripper 1950X with 3.4 GHz and 64 GB RAM. We determine

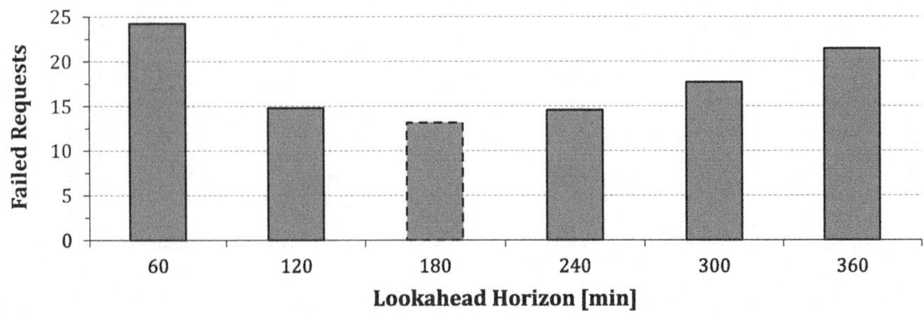

Fig. 8.7 Determining the lookahead horizon of SLA_{on}

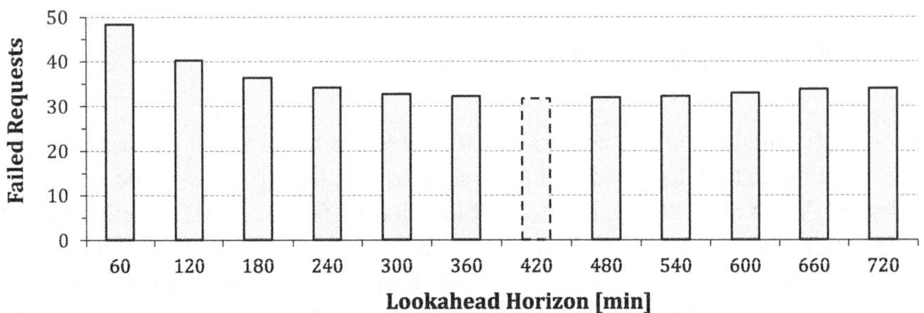

Fig. 8.8 Determining the lookahead horizon of SLA_{off}

dynamic lookahead horizons for LAs with online and offline simulation component, and for vehicle fleets of one to four vehicles. Therefore, we have $2 \cdot 4 = 8$ experimental setups for every BSS. The resulting sequences of lookahead horizons are depicted in Table A.4 of Appendix A.

8.5.5 Rollout Algorithms

In this section, we aim on identifying suitable numbers of simulation runs for the rollout algorithms. The elementary rule is that the solution quality and the runtime as well increase with the number of simulation runs. Further, we follow the idea of LAs and limit the lookahead horizon.

Rollout algorithms using DLA_{off} as base policy draw on the horizons predetermined by the VFA in Sect. 8.5.4. For STR, we use the best safety buffer identified in Sect. 8.5.1. That is 0.1 for Minneapolis and 0.2 for San Francisco. Rollout algorithms using STR as base policy draw on horizons $\delta \in \{60, \dots, 360\}$. Due to the high runtimes, we only apply one vehicle for relocations. Therefore, we end up with $1 + 6 = 7$ experimental setups of rollout algorithms for every BSS. Also due to the runtimes, we are content with 100 instances to evaluate the experimental setups.

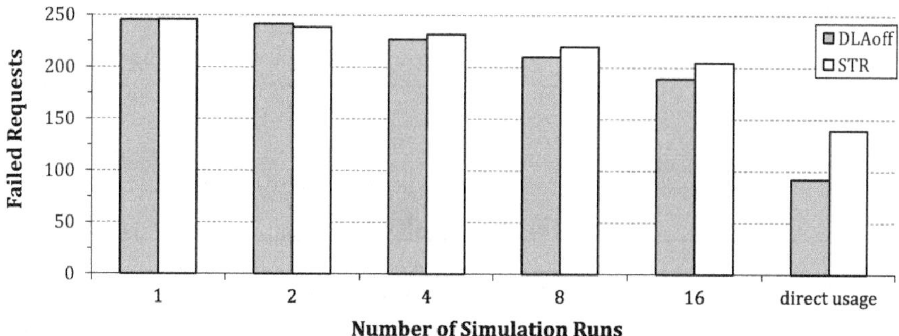

Fig. 8.9 Simulation runs of rollout algorithms in Minneapolis: failed requests

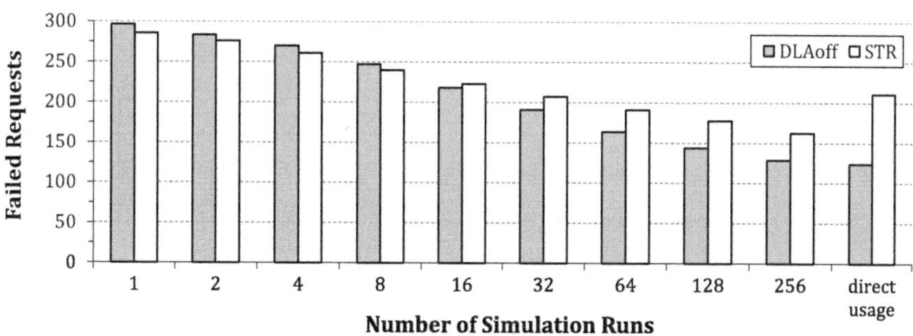

Fig. 8.10 Simulation runs of rollout algorithms in San Francisco: failed requests

Figures 8.9 and 8.10 depict the failed requests in Minneapolis and in San Francisco. We investigate various numbers of simulation runs depicted on the abscissas. Additionally, the last entry on the abscissas refers to the direct usages of the base policies. The resulting failed requests are depicted on the ordinates. The grey bars refer to rollout algorithms using DLA_{off} as base policy. The white bars refer to those applying STR. For STR, the results for the best performing lookahead horizon are shown. We observe, that rolling out STR is advantageous compared to rolling out DLA_{off} if the number of simulation runs is low. From four simulation runs on in Minneapolis and from 16 on in San Francisco, rolling out DLA_{off} is beneficial. For Minneapolis, the maximum number of simulation runs investigated is 16. Until that number, the rollout algorithms are not able to achieve better results than the direct usage of the associated base policy. For San Francisco, the rollout algorithm using STR outperforms the direct usage if at least 32 simulations are conducted. From that number on, the advantage increases. Here, the maximum number of simulation runs investigated is 256.

Figures 8.11 and 8.12 depict the runtimes of rollout algorithms in Minneapolis and in San Francisco. On the ordinates, the runtimes are depicted in hours. The grey bars show the average runtimes of the rollout algorithm using DLA_{off}. The white bars show the average sum of rollout algorithms using STR with the six lookahead

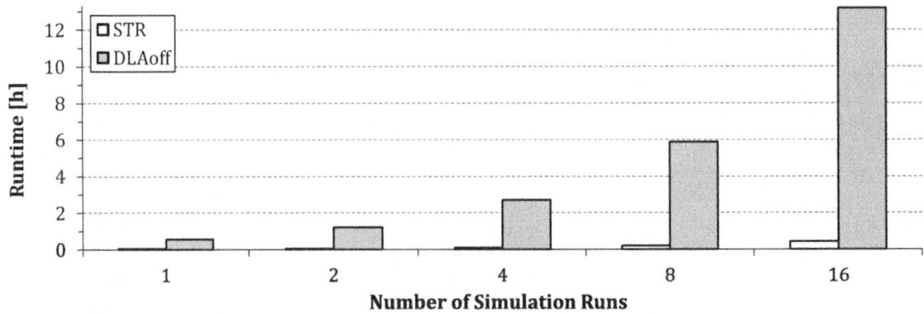

Fig. 8.11 Simulation runs of rollout algorithms in Minneapolis: runtime

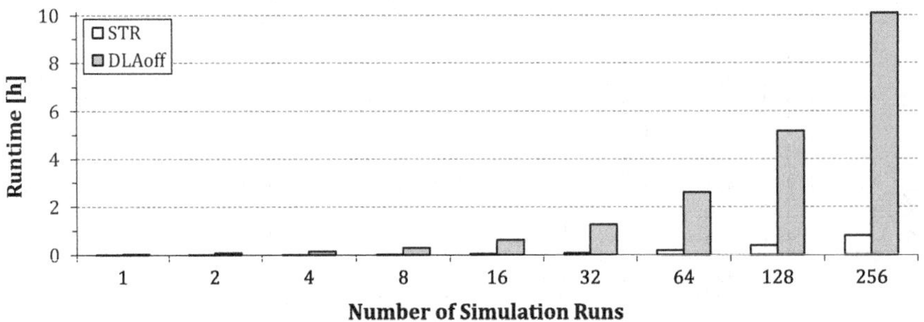

Fig. 8.12 Simulation runs of rollout algorithms in San Francisco: runtime

horizons.[4] The bars point out that the runtime increases with the number of simulation runs. Further, the runtime of the rollout algorithm using DLA_{off} is higher than the sum of runtimes of the rollout algorithms using STR.

Increasing the numbers of simulation runs and, in this way, reducing the numbers of failed requests, would lead to even higher runtimes. Even though the number of test instances is reduced, an evaluation would hardly be possible. Therefore, we conclude that rollout algorithms are not suitable for solving the MDP of the IRP_{BSS}.

8.6 Results

In this section, we present the results achieved by the different policies. For every policy, 1,000 test instances are solved. In Sect. 8.6.1, we investigate the value of coordination by comparing the results of coordinated and independent dispatching. We investigate the value of anticipation by comparing the results of the LAs and STR in Sect. 8.6.2. In Sect. 8.6.3, we present the individual results of our policies.

[4]As in Sect. 8.5.2, we have to investigate all horizons to judge the solution quality. Therefore, we draw on the sum of runtimes.

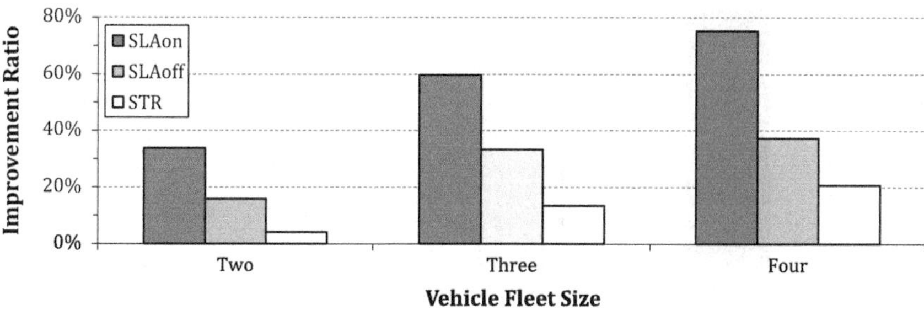

Fig. 8.13 The values of coordination

8.6.1 The Value of Coordination

The value of coordination is depicted by the improvement ratio a policy achieves with coordinated dispatching compared to independent dispatching. This improvement ratio is defined in Eq. (8.6). Let $\mathscr{Q}_\pi, \mathscr{Q}_{\pi_{\text{ind}}} \in \mathbb{R}_0^+$ be the average failed requests achieved by policy π using coordinated dispatching or independent dispatching, respectively.[5] For the LAs, coordinated dispatching is achieved by means of the matrix maximum approach. For STR, we do not send a vehicle to a station if an other vehicle is currently traveling to this station. Then, the improvement ratio $\mathscr{I}_\pi$ depicts the portion of requests a policy π can save by means of coordination.

$$\mathscr{I}_\pi = 1 - \frac{\mathscr{Q}_\pi}{\mathscr{Q}_{\pi_{\text{ind}}}} \tag{8.6}$$

Figure 8.13 depicts the values of coordination. On the abscissa, the vehicle fleet sizes of two, three, and four are separated. On the ordinate, the improvement ratios are indicated. The bars depict the average improvement ratios of the associated policy using the vehicle fleet size stated over both BSSs. The dark grey bars refer to SLA_{on}, the light grey bars to SLA_{off}, and the white bars to STR. We can see that every policy takes advantages if coordinated dispatching is applied. Further, we observe two trends. First, the more vehicles are applied, the higher the improvement ratio is. Second, SLA_{on} takes most advantages, SLA_{off} takes second most, and STR takes least advantage.

8.6.2 The Value of Anticipation

The value of anticipation is depicted by the improvement ratio a policy achieves by applying simulations. This improvement ratio is defined in Eq. (8.7). Here, let

[5]No policy achieves average failed requests equal to zero in any experimental setup.

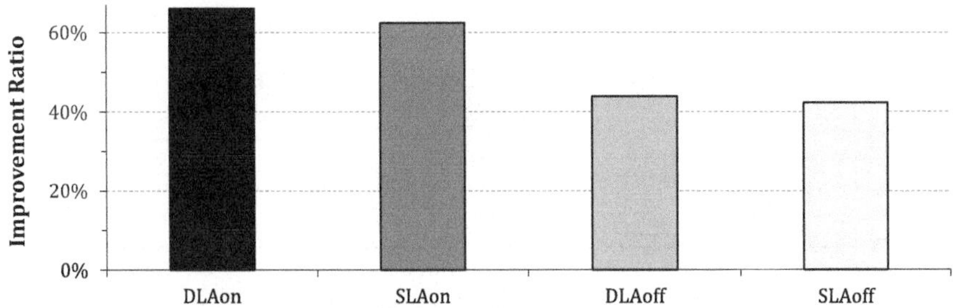

Fig. 8.14 The values of anticipation

$\mathcal{Q}_\pi \in \mathbb{R}_0^+$ be the average failed requests achieved by a policy π. We compare the failed requests achieved by the LAs and STR. Thus, the improvement ratio $\mathcal{I}_\pi$ here depicts the portion of requests a policy π can save by means of simulation.

$$\mathcal{I}_\pi = 1 - \frac{\mathcal{Q}_\pi}{\mathcal{Q}_{\pi_{\mathrm{STR}}}} \tag{8.7}$$

In Fig. 8.14, the average improvement ratios of the different LAs are shown. On the abscissa, the policies are separated. The bars depict the average improvement ratios of the associated policy over vehicle fleet sizes from one to four and both BSSs. The black bar refers to $\mathrm{DLA}_{\mathrm{on}}$, the dark grey bar to $\mathrm{SLA}_{\mathrm{on}}$, the intermediate grey bar to $\mathrm{DLA}_{\mathrm{off}}$, and the light grey bar to $\mathrm{SLA}_{\mathrm{off}}$. The bars are depicted in descending order according to their average improvement ratios. We can see that the LAs with online simulation component achieve improvement ratios of more than 60%. The LAs with offline simulation component outperform the associated STR by more than 40% on average. From this point of view, the LA take huge advantages from online simulations and little advantages from dynamic lookahead horizons.

8.6.3 Individual Results

The individually failed requests achieved by the LAs and by vehicle fleet sizes of one to four in Minneapolis and in San Francisco are illustrated by Figs. 8.15 and 8.16. The vehicle fleet sizes are shown on the abscissas. The ordinates depict the failed requests. The black bars refer to $\mathrm{DLA}_{\mathrm{on}}$, the dark grey bars to $\mathrm{SLA}_{\mathrm{on}}$, the intermediate grey bars to $\mathrm{DLA}_{\mathrm{off}}$, and the light grey bars to $\mathrm{SLA}_{\mathrm{off}}$.

The figures depict that the general level of failed requests decreases if more vehicles are applied. Moreover, the more vehicles are applied, the more advantage can be taken from using the online simulation component. This is due to the station interactions included in the online simulations only. In this way, future failed requests are approximated more precise and the anticipation of the fleet's future decisions by means of coordination is supported. The more vehicles applied, the more important coordination is.

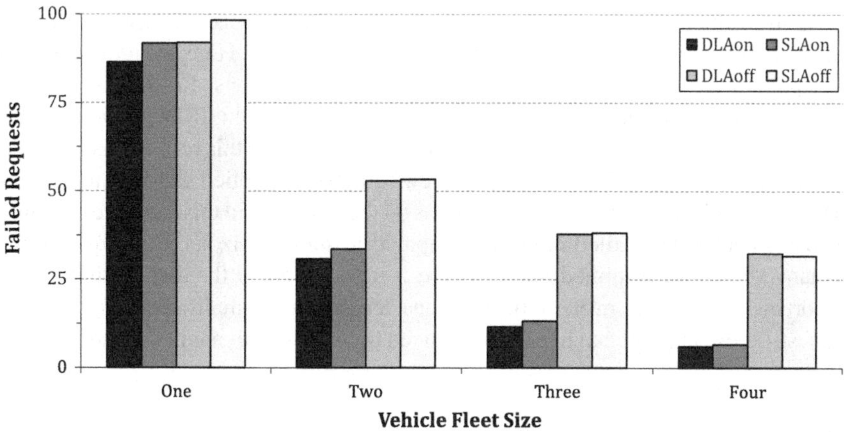

Fig. 8.15 The results of Minneapolis

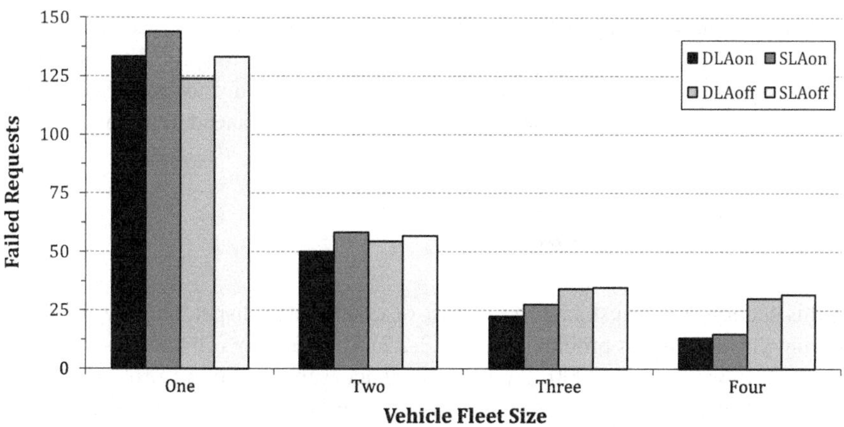

Fig. 8.16 The results of San Francisco

Apart from one exception, the LAs using online simulations are always beneficial compared to those with offline simulation component. Additionally, in Minneapolis the difference between LAs with online and offline simulation components is much larger than in San Francisco. Only in the case of one vehicle in San Francisco, DLA_{off} and SLA_{off} outperform DLA_{on} and SLA_{on}. As described in Chap. 6, the online simulation component promises a higher solution quality compared to the offline simulation component. However, due to the dominating commuter usage, we assume the request pattern in San Francisco to be more regular than in Minneapolis. In contrast, due to the significant leisure activities, we assume more unexpected requests in Minneapolis. As the offline simulation component exclusively draws on average requests, the associated LAs relatively perform better in San Francisco.

Again with one exception, the LAs using dynamic lookahead horizons are always superior compared to their counterparts using static horizons. Here, the more vehicles are applied, the lower the advantage from dynamic horizons. In the case of four vehicles and Minneapolis, SLA_{off} outperforms DLA_{off}. In the IRP_{BSS}, the vehicles are the resource to minimize the amount of failed requests. If this resource is scarce, it has to be managed well. In other words, if few vehicles are applied, the dispatching has to be done well. As the dispatching depends on the horizons, in this case, the dynamic horizons lead to less failed requests compared to static horizons. Consequentially, if many vehicles are applied, the resource is not scarce and the dispatching is less important. Further, as even static horizons lead to very small amounts of failed requests, these amount can hardly be reduced by means of dynamic horizons.

8.7 Analysis

In this section, we analyze the results. In Sect. 8.7.1, we investigate the results achieved by means of solving the coordination problem's ILP to optimality. The failed requests in the course of a VFA's approximation phase are discussed in Sect. 8.7.2. In Sect. 8.7.3, we investigate the structure of dynamic lookahead horizons.

8.7.1 Optimal Assignment

In this section, we investigate the impact of coordinated dispatching by means of solving the assignment problem (Sect. 6.2.2.2) to optimality. This analysis is motivated by Powell et al. (2000) studying a related vehicle routing problem. In the referenced work, vehicles are assigned to customers. Uncertainty is given as customers request stochastically when the vehicles are already on the road. Whenever new customers request, vehicles are dynamically redispatched by means of solving an assignment problem. The assignment problem is deterministic since it includes the information that are certain at the point in time it arises. Solving the assignment problem heuristically leads to advantages compared to an optimal assignment. Due to reasons discussed in the following, we detect similar results when solving our assignment problem in the LAs' optimization component either heuristically or to optimality.

The assignment problem uses the failed requests approximated in the simulation component of the LA. In the optimization component, the approximations are used to determine the requests a certain vehicle can save at a certain station. It is clear that the requests used in the simulation component will not occur in the actual MDP to solve. Therefore, the assignment problem's optimal solution can hardly be optimal in the MDP to solve. Further, and speaking generally, particular structures in an instance are always reflected by structures in an optimal solution. Therefore, optimal solutions are most likely not flexible if applied for another instance. For the assignment problem

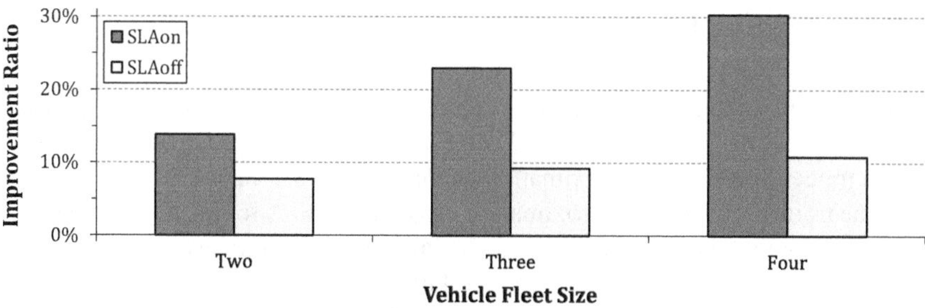

Fig. 8.17 Improvement ratios of heuristic over optimal assignment

investigated here, a heuristic assignment provides the required flexibility to take advantage over an optimal assignment if the requests in the actual MDP are not equal to those in the simulation component.

We consider the average failed requests $\mathscr{Q}_\pi, \mathscr{Q}_{\pi_{\text{opt}}} \in \mathbb{R}_0^+$ achieved by a policy π by means of solving the assignment problem either heuristically or to optimality. Therefore, we draw on the matrix maximum approach and on the Hungarian algorithm (Bazaar et al. 2010). The Hungarian algorithm solves the assignment problem to optimality within polynomial time.[6] We determine the improvement ratio $\mathscr{I}_\pi$, i.e., the portion of saved requests by using the heuristic instead of the optimal assignment, according to Eq. (8.8).

$$\mathscr{I}_\pi = 1 - \frac{\mathscr{Q}_\pi}{\mathscr{Q}_{\pi_{\text{opt}}}} \tag{8.8}$$

We determine the improvement ratios for LA_{on} and LA_{off}. To this end, we conduct experiments with lookahead horizons $\delta \in \{60, \dots, 360\}$ minutes for LA_{on} and $\delta \in \{60, \dots, 720\}$ minutes for LA_{off}. Fleets of two, three, and four vehicles are applied. Therefore, we conduct $(6 + 12) \cdot 3 = 54$ experiments for every BSS. The individual results for every experiment are depicted in Appendix B. For the analysis here, we use the best performing lookahead horizon for every fleet and BSS. The improvement ratios are depicted in Fig. 8.17. On the abscissa, the vehicle fleet sizes are shown. The ordinate indicates the improvement ratios. The bars refer to the average improvements ratios for the respective vehicle fleet size over both BSSs. The dark grey bars refer to LA_{on} and the light grey bars to LA_{off}. We can see that for every vehicle fleet size, the LAs can improve their solution quality by means of heuristic assignments. The more vehicles are applied, the larger the advantage. Further, LA_{on} takes more advantage than LA_{off}.

[6]We draw on an implementation provided by Stern (2012).

8.7.2 Learning Curves

In this section, we analyze the iterative improvements achieved in the approximation phase of the DLAs' VFA. To this end, we examine the average failed requests over certain trajectories in the approximation phase. In Figs. 8.18 and 8.19, the abscissas show the trajectories of the approximation phase of the VFA for DLA_{on}, one vehicle, and Minneapolis. The ordinates indicate the average failed requests.

The learning curves depict the average failed requests until a certain trajectory. Figure 8.18 depicts the first 500 trajectories whereas Fig. 8.19 depicts the whole approximation phase until the termination criterion occurs. Especially in Fig. 8.18, we can see the curve jittering. The jittering is due to the Boltzmann exploration more or less randomly selecting lookahead horizons. When a reasonable number of trajectories has been traversed and a reasonable number of observations is included in the values, the procedure turns more to exploitation. Then, in Fig. 8.18, the curve indicates a smooth progress. The progress depicts a constant improvement. In trajectories 33,649–34,649, the Boltzmann exploration always selects the horizons with minimal values. Therefore, the stop criterion holds and the approximation phase is stopped.

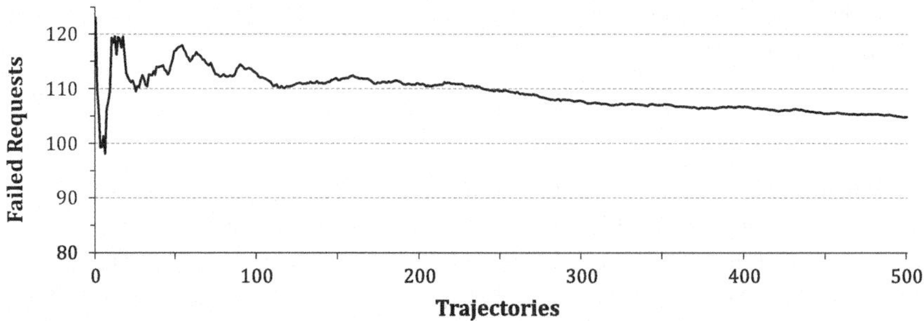

Fig. 8.18 Failed requests in the first 500 trajectories of an approximation phase

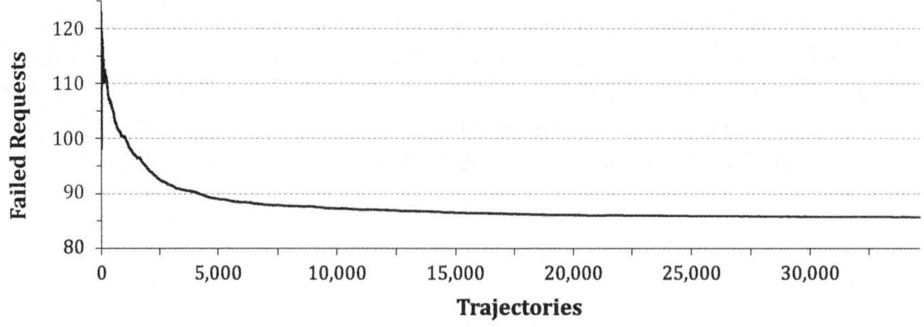

Fig. 8.19 Failed requests in the course of an approximation phase

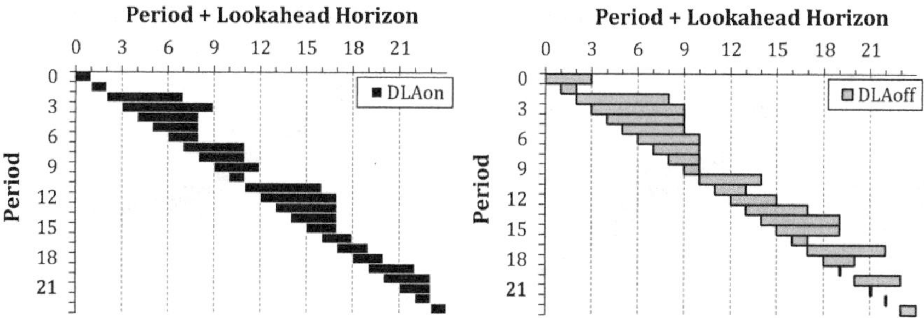

Fig. 8.20 VFA-determined lookahead horizons for one vehicle in Minneapolis and San Francisco

8.7.3 *Dynamic Lookahead Horizons*

In this section, we analyze the DLAs' lookahead horizons. To this end, we discuss the dynamic lookahead horizons determined by the VFA in Sect. 8.7.3.1. As we select horizons for periods (and neglect spatio attributes), we try to explain the horizons with the aid of the temporal pattern of requests. We will see that this is partially possible. Additionally, in Sect. 8.7.3.2, we approach dynamic lookahead horizons determined manually. We already saw that the improvement achieved by dynamic horizons reflects the availability of vehicles (Sect. 8.6.3). Therefore, in the following two sections, we exemplarily show the lookahead horizons for the cases of one and four vehicles.

8.7.3.1 VFA-Determined Lookahead Horizons

For one vehicle, the LAs take most advantage from dynamic lookahead horizons. Here, the structures of the horizons reflect the request patterns. For four vehicles, the LAs failed to take advantage in Minneapolis and take little advantage in San Francisco (Sect. 8.6.3). In this cases, the horizons do not reflect the request patterns.

Figures 8.20 and 8.21 exemplary illustrate the lookahead horizons in the overall time horizon for one and four vehicles.[7] The ordinates show the periods, i.e., the hours of the day. The abscissas show the periods and the length of the associated horizon. The bars indicate the periods covered by the DLAs' simulations.

We first analyze the case when of applying one vehicle. On the left-hand side of Fig. 8.20, the lookahead horizons of DLA_{on} for one vehicle in Minneapolis are shown. The right-hand side shows the horizons of DLA_{off} for one vehicle in San Francisco. In both BSSs, the amounts of trips have peaks in hours 8, 12, and 17 (Fig. 8.1, p. 83). We can see that in Minneapolis, the lookahead horizons reflect the three peaks. Regarding Minneapolis, in the morning, the horizons extend until hours 7–9. Before noon, the horizons reach until hours 11–12, and in the afternoon until hours 16–18.

[7] All VFA-determined lookahead horizons can be looked up in Table A.4 (Appendix A).

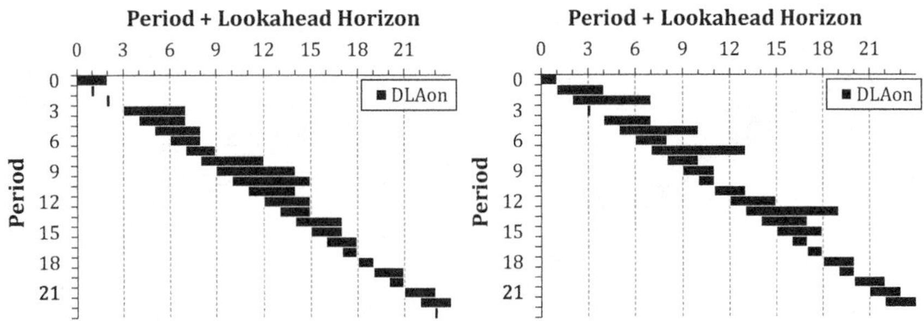

Fig. 8.21 VFA-determined lookahead horizons for four vehicle in Minneapolis and San Francisco

According to this structure, we can conclude that simulations beyond peak hours is not beneficial. Then, the vehicle conducts relocations in order to prepare the BSS for the next peak. In San Francisco, the horizons reflect the morning peak. Here, the horizons extend until hours 7–10. A possible explanation for these different structures is that in Minneapolis the second peak is higher than the first one, and the third peak is higher than the second one. Further, the second and third peaks probably are also due to leisure activities. The resulting request require the vehicles to prepare the BSS for this purpose. The lookahead horizons as determined by VFA allow to explore the pattern. In San Francisco, the horizons reflect the first peak only. Here, the first peak is higher than the other two and, most likely, commuter usage is dominating all day. Therefore, the BSS is prepared once for this purpose. After the first peak, we cannot recognize any clear structure.

In the following, we analyze the case of a fleet of four vehicles. Figure 8.21 shows the lookahead horizons DLA_{on} for Minneapolis on the left-hand side and for San Francisco on the right-hand side. For Minneapolis, we see a more or less regular structure. That is, the horizons mostly have a length of one to three hours. In hours 8–13, the horizons reach until hour 15. This can hardly be explained by the temporal distribution of trips. In the horizons of San Francisco, no structure can be observed.

8.7.3.2 Manual-Determined Lookahead Horizons

In this section, we analyze manually determined lookahead horizons and the corresponding results. Therefore, we introduce the A Priori DLAs and Ex Post DLAs.

A Priori DLA

For the A Priori DLA, we manually adapt the horizons to the temporal distribution of request. That is, we simulate until the next peak of request. This idea is motivates by the findings of the VFA-determined horizons of DLA_{on} for one vehicle in Minneapolis (Fig. 8.20). Figure 8.22 shows the resulting horizons. The horizons reach until the request peaks in hours 8, 12, 17, and until the end of the day. These horizons are used

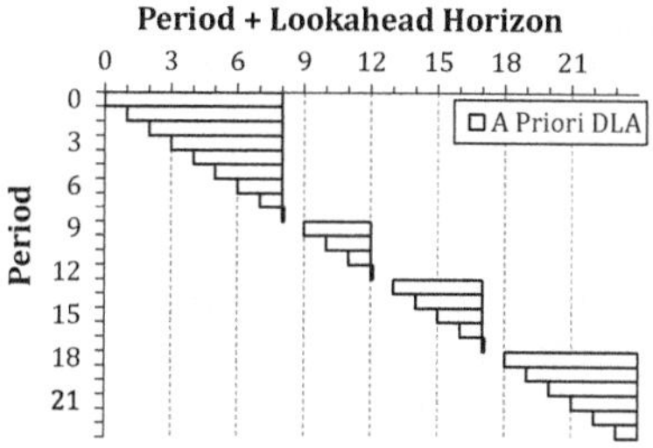

Fig. 8.22 A priori DLAs' lookahead horizons

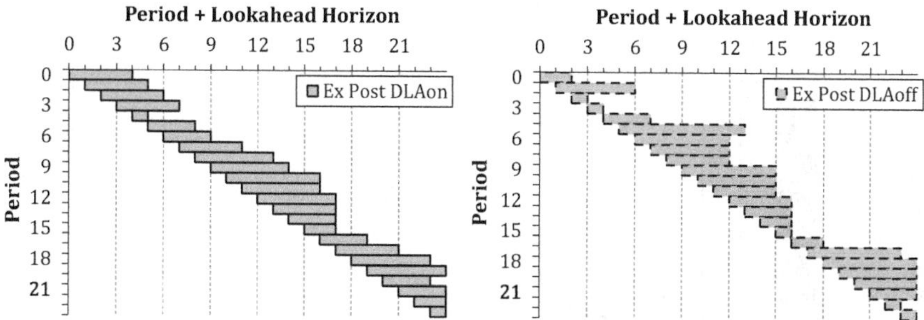

Fig. 8.23 Ex Post DLAs' lookahead horizons one vehicle in Minneapolis and San Francisco

for every vehicle fleet size and for both BSSs. Further, we apply online and offline simulations.

Ex Post DLA

The basic idea of the Ex Post DLA is to recall the results of the SLAs with various horizons (Sect. 8.5.3). We allow the fallacy that the selection of a lookahead horizon is responsible for the failed request in this certain current period. To this end, in every period, we manually select the horizon of the SLA that performed best in the associated period. Figures 8.23 and 8.24 exemplary show lookahead horizons for one and four vehicles.[8] In advance of the results, we analyze the horizons and results of the Ex Post DLAs with best performing simulation component. The bars in the following figures are solid-framed if they refer to Ex Post DLAs with online simulation component. Bars referring to Ex Post DLAs with offline simulations are dashed framed. Figure 8.23 shows the horizons of Ex Post DLAs for one vehicle in Minneapolis on the left-hand side and in San Francisco on the right-hand side. The case of one vehicle in Minneapolis is characterized by a certain regularity. Only the afternoon peak is reflected by the horizons. The case of one vehicle in San Francisco the noon and afternoon peaks are visible. Figure 8.24 shows the corresponding horizons for four vehicles. Notably, the structures of horizons for Minneapolis (left) and

[8] All manual-determined lookahead horizons can be looked up in Table A.5 (Appendix A).

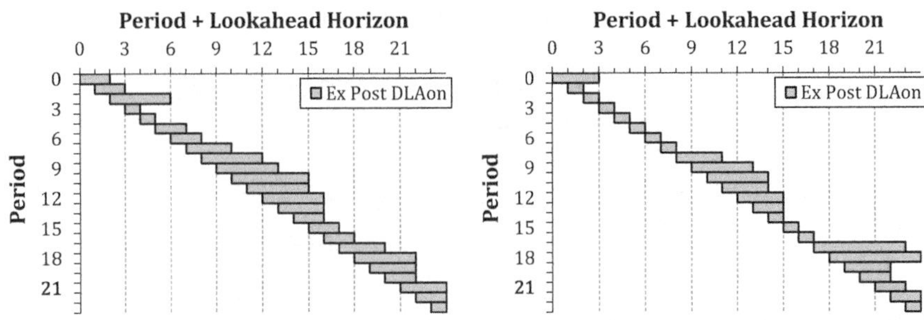

Fig. 8.24 Ex Post DLAs' lookahead horizons for four vehicles in Minneapolis and San Francisco

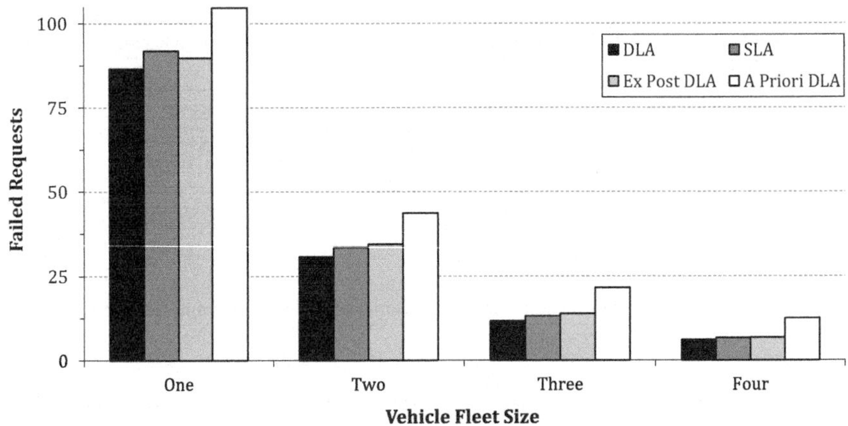

Fig. 8.25 The results of DLAs with manual-determined horizons in Minneapolis

San Francisco (right) reveal significantly shorter horizons than for four vehicles. In the case of four vehicles in Minneapolis the afternoon peak is reflected.

In the following, we analyze the results achieved by the DLAs applying VFA-determined and manual-determined horizons. Additionally, we involve the results of the best performing SLAs. For Minneapolis, the LAs with online simulation component outperform those applying offline simulations. For San Francisco, in some cases the offline simulations are beneficial.

In Figs. 8.25 and 8.26, the results for Minneapolis and San Francisco are shown.[9] On the abscissas, the vehicle fleet sizes are shown. The ordinates show the failed requests. A bar is dashed-framed if in the associated case the offline simulations are beneficial. Black bars refer to VFA-based DLAs, dark grey bars to SLAs, light grey bars to Ex Post DLAs, and white bars show the results of A Priori DLAs. Once more, the results point out, that the general level of failed requests is decreased if more vehicles are applied. However, the DLAs with manual-determined horizons are inferior compared to DLAs with VFA-determined horizons. Except of one case,

[9]The numerical results are shown in Appendix B.

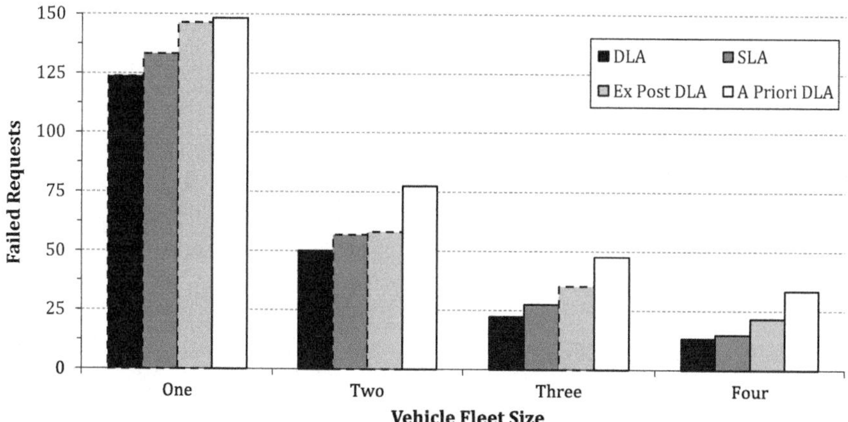

Fig. 8.26 The results of DLAs with manual-determined horizons in San Francisco

also the SLAs outperform the DLAs with manually determined horizons. In the case of Minneapolis and one vehicle, the Ex Post DLA achieves less failed requests than SLA. Additionally, we observe that the A Priori DLAs are inferior compared to the Ex Post DLAs. We can conclude that applying an VFA to determine dynamic lookahead horizons is beneficial.

Part III
Conclusion

Chapter 9
Managerial Implications

In this chapter, we summarize the findings of the previous chapters to accentuate managerial implications.

The results of the case studies point out that our approaches lead to a significantly higher solution quality compared to benchmark approaches from literature. More precisely, we successfully integrate data in a dynamic decision process in order to approximate and anticipate future events. Approximations are obtained from simulations over a limited lookahead horizon. Anticipation is achieved by including the approximations in the decision making process.

Our approaches outperform the benchmarks by far if requests follow a clear structure as well as if a significant amount of requests is hard to explain. However, the solution quality is relatively higher in the latter case as our approach explicitly accounts for active decision making.

Even though, the vehicles are individually dispatched, coordination of the vehicle fleet is highly beneficial especially if many vehicles are applied. That is, we anticipate possible decisions of the whole fleet when making decisions for one vehicle. The first resulting benefit is that the fleet minimizes the amount of saved request. Second, we preserve a certain flexibility in order to counteract unexpected requests. Further, the anticipation is supported by means of including station interactions in the simulations.

We identify stations where relocations are most urgent by limiting the lookahead horizon. Due to the spatio-temporal distribution of requests, suitable horizons vary in the course of the day. Therefore, we apply dynamic horizons selected for certain periods of the day. Dynamic horizons achieve improvements of the solution quality especially if few vehicles are applied.

Although the approaches performed well in the experiments of the case studies, we must state that anticipation in practice strongly depends on the availability as well as on the quality of data. We have to admit that in the case study, the data quality may be impaired as we used data of successful request only. This is a shortage if the distribution of successful and failed requests is not the same. To the best of our knowledge, data on failed requests is not available as observing failed requests is challenging in practice. When users arrive at a station that cannot serve their request, they probably will not report the failed request. Instead, they will immediately look

J. Brinkmann, *Active Balancing of Bike Sharing Systems*, Lecture Notes in Mobility,
https://doi.org/10.1007/978-3-030-35012-3_9

for an alternative station. Nevertheless, in some BSSs, users are granted additionally time for their trip when they request to return a bike at a full station (Vogel 2016). To this end, the stations' terminals have a corresponding feature. However, failed rental requests remain unobserved. Therefore, we advise providers to implement incentives motivating users to report both failed rental and return requests, and to publish the resulting data. We argue that information gained in such a way will significantly enhance the quality of data and, therefore, the applicability of formal optimization approaches in practice.

Chapter 10
Future Research

In this chapter, we point out future research. In Sect. 10.1, we suggest possible extensions of the model to account for challenges in practice. Possible extensions of the method to obtain a higher solution quality are suggested in Sect. 10.2.

10.1 Model

We suggest five extensions of the model addressing aspects of fairness, uncertain travel times, moveable stations, the integration of social media, and battery chargers.

Fairness

The first suggestion for a model extension addresses fairness aspects. In the IRP_{BSS}, it may happen that stations are not visited by vehicles although they are empty or full. However, some BSS providers are instructed by local administrations to leave stations empty of full for a maximum time span only (Vogel 2016). Otherwise, some imbalanced stations would likely never be visited. Therefore, providers need to account for fairness. Soeffker et al. (2017) study a related VRP and discuss fairness aspects. In the VRP, customers stochastically request a service over a finite time horizon. Every customer's location is unknown to the dispatcher until the customer requests. Whenever a customer requests, one either has to accept or reject the request. Accepted requests are served by one uncapacitated vehicle. Therefore, the vehicle's route is dynamically adapted whenever a request is accepted. The locations are uniformly distributed in a Euclidean service area. A depot is located in the center of the area. The vehicle starts and ends its tour at the depot. The objective is to maximize the number of served customers. Methods of ADP are applied to solve the VRP. An analysis of the solution structure reveals that the customer location has a significant impact on whether a request is accepted or not. To be precise, the closer a customer is

J. Brinkmann, *Active Balancing of Bike Sharing Systems*, Lecture Notes in Mobility,
https://doi.org/10.1007/978-3-030-35012-3_10

located to the depot, the higher the probability to be accepted. In practice, this leads to dissatisfaction, or in other words, to unfairness. In the IRP_{BSS}, we can counteract unfairness by means of using additional penalties for failed requests at stations at the brink of the BSS's operation area or at stations that have been full or empty for a predefined time. These additional penalties can be anticipated by the LAs and, in this way, fairness will be achieved.

Travel Times

The second suggestion addresses uncertain travel times. Especially in urban areas, travel times are a significant driver of uncertainty (Ulmer 2017). In the LAs, travel times are determining the amount of requests a vehicles can save at a station (Sect. 6.2.2.2). Therefore, the approximations of saved requests are as accurate as the travel time predictions. Groß et al. (2017) introduce an approach to account for reliability in travel times. They obtain travel times of street segments from data recorded in Winnipeg (Canada) and a traffic simulation. In a case study, they evaluate paths between varying origins and destinations and in different periods of a day. A reliability criterion is defined to estimate how accurate predictions of the travel time can be made. The authors claim that the usage of reliable paths is advantageous in the presence of uncertain travel time compared to paths obtained from conventional shortest path algorithms. Köster et al. (2018) introduce an approach accounting for sudden travel time changes on certain street segments. Such changes are due to dynamic traffic management systems. Such systems may change the speed limit or the direction of certain road lanes on short notice. This impacts the road capacities and, therefore, the travel times. The approach by Köster et al. (2018) anticipates sudden travel time changes. The approach is evaluated in a case study covering a VRP where a set of customers needs to be visited. The objective is to route a vehicle in the presence of uncertain travel times such that the tour duration is minimized. The instances used are obtained from real-world data by Braunschweig (Germany). The results of the case study reveal that the approach anticipating travel times is beneficial compared to approaches drawing on constant travel times. The approaches by Groß et al. (2017), Köster et al. (2018) can be integrated in the IRP_{BSS} by means of extending the travel time function by a time component. Then, the associated path's travel time is returned for a given pair of station at a given point in time. In this, way, the saved requests can precisely be approximated.

Moveable Stations

The third suggestion is motivated by Manna (2016) allowing a certain flexibility of station capacities. Manna (2016) studies a problem where so called moveable stations are used for a temporal extension of stations. A moveable station is placed alongside a station and provides additional bike racks to support the balancing of the BSS. The moveable stations are hauled by transport vehicles. This concept can be integrated as an additional feature of the IRP_{BSS} by means of additional state attributes and decision variables. However, we further suggest to combine the approach by Manna (2016) with an approach on autonomous vehicles (AVs). Scherr et al. (2018) study a problem on AVs delivering parcels within an urban area. The AVs are allowed to move autonomously in certain AV zones. To move outside this zones, the AVs are attached to a manned vehicle. We claim that AVs can be used as moveable station in the sense of Manna (2016). AVs need to be set up so that they can carry bikes. Then, they can autonomously move to a station in order to extend the station's capacity. During the day, the AVs can change their stations in order to counteract sudden imbalances. The advantage of using AVs as moveable stations in addition to conventional and manned vehicles is that more stations can be served at the same time without additional staff.

Social Media

The fourth suggestion addresses the integration of social media to support the balancing of BSSs. As stated in the previous chapter, the quality of the data on the request pattern is possibly impaired as observing failed requests is challenging in practice. However, users likely complain about frequently imbalanced stations or point out possible locations for new stations in social media. Deriyenko et al. (2017) introduce an approach to extract customer feedback from social media and to make it valuable for decision makers. Therefore, we suggest to use the approach by Deriyenko et al. (2017) to enhance the data quality.

Battery Chargers

The fifth suggestion accounts for electric mobility and the need for battery chargers. We claim, that our approach basically suits for systems offering pedelecs, i.e., electrically driven bikes and/or e-scooters. Though, in both cases, the stations need to be equipped with devices to charge the batteries. If a user returns a pedelec or an e-scooter with a low state of charge, it is not available for other users immediately. Instead, it needs to be blocked for a certain time to charge its battery. Therefore, from the model perspective, the term fill level becomes a new meaning. In the application of electric mobility, the fill level describes the number of *available* pedelecs or e-scooters. After a certain time, they automatically become available for rentals again and, therefore, the fill level increases automatically.

10.2 Method

The five extensions of the method address risk aversion, state space partitioning, spatial attributes in the VFA, master tours, and dynamic policy selection.

Risk Aversion

The first suggestion regarding an extension of the method accounts for risk aversion. In the objective function of the IRP_{BSS} as well as stochastic-dynamic programs in general, the expected contributions are considered only. Therefore, the decision maker is implicitly assumed to be risk neutral. However, also the contributions' variances are relevant when taking decisions (Markowitz 1952). To account for risk aversion, Ulmer and Voß (2016) include a risk measure in the decision making process. The authors investigate a VRP where customers request service over a finite horizon. The actual decisions are to accept or to reject the customers. Accepted customers need to be visited by one vehicle. The objective is to maximize the amount of served customers. The authors introduce a VFA to approximate the values of states and decisions as well as an associated risk measure. When taking a decision, the values and the risk measures are weighted by means of a predefined factor representing the degree of the dispatcher's risk aversion. By doing this, decisions with an uncertain outcome are not taken if the dispatcher is risk-averse. The approach by Ulmer and Voß (2016) can be integrated in the LAs by means of considering the variance of failed requests observed in the online lookahead component.

State Space Partitioning

The second suggestion addresses the state space partitioning within the LUT of the VFA. In the VFA, we use twenty-four periods each of one hour length. This partitioning may be natural for human beings but – as a matter of facts – it is chosen arbitrarily. Soeffker et al. (2016), Ulmer et al. (2017b) introduce approaches for dynamic state space partitioning in a VFA-based method to solve a VRP. To this end, the authors of the two studies draw on cluster analysis. Soeffker et al. (2016) use a predefined number of representatives located in the state space. Then, a partition-based clustering algorithm assigns states to representatives. The location of the representatives and the assignment of states are iteratively updated. Ulmer et al. (2017b) partition the state space by means of hierarchical clustering. Both Soeffker et al. (2016), Ulmer et al. (2017b) compare their approaches with predefined and static state space partitioning. It turns out that dynamic state space partitioning is beneficial in terms of solution quality. Additionally, Ulmer et al. (2017b) emphasize the efficiency of dynamic state space partitioning in terms of the approximation phase's run time. Especially the approach by Ulmer et al. (2017b) can be used to identify suitable periods for the state space partitioning used by the VFA we apply.

Spatial Attributes

The third suggestion addresses the consideration of spatial attributes for the state space partitioning. So far, we used the point in time only. According to a study by Ansmann et al. (2017), spatial attributes are also promising. Ansmann et al. (2017) study a VRP where customers' requests are accepted for service or rejected. Accepted customers are visited by a vehicle. The objective is to accept as many customers as possible. When deciding about accepting or rejecting a request, primarily the resulting vehicle detour is considered. However, the customer location supports routing the vehicle towards fruitful regions, i.e., where future requests are expected. The integration of the spatial attribute results in an higher solution quality as well as in higher runtimes of the approximation phase. Spatial attributes can also be used for the state space partitioning of our VFA. The VFA we use selects lookahead horizons for all stations with respect to the period only. It may be beneficial to vary the LA's lookahead horizon also with respect to the districts the stations are located.

Master Tours

The fourth suggestion accounts for the portion of requests that can be forecasted close to certainty. For instance, in the presence of a major public event, there is most likely no advanced data analysis or simulation required to predict full stations near the venue before the event starts and to predict empty stations as soon as the event ends. Therefore, the vehicles of the BSS in Paris (France) follow different balancing strategies. The majority of vehicles is dynamically dispatched, but a few vehicles visit stations along master tours (Legros 2019). Master tours are tours that are predetermined before the vehicles are on the road. Neumann Saavedra et al. (2016) determine master tours with respect to expected requests. Therefore, we suggest to partition the set of stations such that the first subset is served by master tours and the second by means of active relocations. The partitioning may be adapted with respect to different periods of the day. Further, master tours may also be beneficial when the user activity is low, e.g., at night.

Dynamic Policy Selection

The fifth suggestion aims on an advanced usage of the VFA. In this work, we use the VFA to select parameters for a well defined policy dynamically. Hermanns et al.

(2019) extend the VFA to realize a dynamic policy selection (DPS). In every decision state of a stochastic-dynamic optimization problem, DPS selects a policy out of a set of unique candidate policies. The advantages of DPS are demonstrated in a case study on a stochastic-dynamic knapsack problem. In every decision state of the knapsack problem, one has to accept or reject an items. Every item has a reward and a weight. Weight and reward are subject to stochastic distributions. However, future items are unknown. The sum of weights over all accepted items must not exceed a given capacity. The objective is to maximize the sum of rewards. Hermanns et al. (2019) define three candidate policies: A greedy policy accepts every item until the capacity is exhausted, an intermediate policy accepts every item where the ratio of reward and weight exceeds a certain threshold, and a policy explicitly anticipating future items. The results point out that DPS significantly improves the solution quality compared to the individual usage of the candidate policies. Regarding the IRP_{BSS}, it may be beneficial to use DPS to identify periods when to use master tours and when to use active relocations.

Appendix A
Parameters

Table A.1 depicts the best lookahead horizons for SLA_{on} and SLA_{off}, and the best safety buffers for STR that have been identified in the preliminary experiments described in Sect. 8.5. We distinguish independent dispatching ("Indep.") and coordinated dispatching by means of heuristic ("Heur.") and optimal assignments ("Opt.").
Unless stated otherwise, we are using these parameters.

Tables A.2 and A.3 show the average failed requests ("Fail.") and runtimes for an entire MDP ("Run.") achieved by SLA_{on} with different horizons and different vehicle fleet sizes in Minneapolis and San Francisco. The runtimes are measured in seconds. The associated experiments are described in Sect. 8.5.2.

Table A.4 shows the horizons for DLAs obtained from the VFA. Table A.5 shows the manually determined horizons. The associated experiments are described in Sects. 8.5.4 and 8.7.3.2.

Tables A.6 and A.7 show the average failed requests ("Fail.") and runtimes for an entire MDP ("Run.") achieved by Rollout algorithms with different base policies. For STR, we investigate simulation horizons from 60 to 360. The runtimes are measured in hours. The associated experiments are described in Sect. 8.5.5.

Table A.1 Parametrization for STR, SLA_{on}, and SLA_{off}

Vehicles		Minneapolis			San Francisco		
		STR	SLA_{on}	SLA_{off}	STR	SLA_{on}	SLA_{off}
	1	0.1	240	240	0.2	300	300
Indep.	2	0.2	180	240	0.2	180	300
	3	0.2	180	240	0.3	120	300
	4	0.2	120	240	0.3	120	240
Heur.	2	0.2	240	240	0.2	300	360
	3	0.2	180	300	0.3	180	300
	4	0.3	180	420	0.3	240	300
Opt.	2	–	180	180	–	240	300
	3	–	180	240	–	240	300
	4	–	180	300	–	180	300

J. Brinkmann, *Active Balancing of Bike Sharing Systems*, Lecture Notes in Mobility,
https://doi.org/10.1007/978-3-030-35012-3

Table A.2 Simulation runs for Minneapolis

Horizon			Simulation runs													
			1		2		4		8		16		32		64	
			Fail.	Run.	Fail.	Run.	Fail.	Run.	Fail.	Run.	Fail.	Run.	Fail.	Run.	Fail.	Run.
Vehicle fleet size	One	60	149.28	0.55	143.57	0.85	138.76	1.45	136.04	2.57	133.61	4.79	131.74	9.27	129.51	17.90
		120	133.50	0.66	121.73	1.01	114.15	1.75	110.82	3.10	108.89	5.77	107.78	11.35	106.37	21.93
		180	123.44	0.84	110.39	1.29	102.22	2.30	98.99	4.20	96.38	8.25	94.72	17.14	93.32	34.47
		240	120.24	1.09	105.88	1.69	98.42	3.19	94.62	6.40	92.77	13.79	91.76	29.60	91.07	61.50
		300	118.67	1.41	104.77	2.21	97.69	4.47	94.72	9.88	93.07	22.50	92.49	50.19	92.79	104.80
		360	120.07	1.78	106.17	2.87	99.56	6.02	96.72	14.03	96.39	33.94	97.07	76.98	97.37	162.80
	Two	60	83.18	1.12	73.45	1.81	67.36	3.27	62.30	6.00	59.26	11.45	57.34	22.72	55.14	43.73
		120	67.92	1.50	55.08	2.40	47.29	4.39	42.09	8.03	39.54	15.31	38.23	30.48	37.08	59.26
		180	62.76	2.02	50.13	3.25	41.64	6.09	37.30	11.44	34.69	22.44	33.80	45.44	31.86	90.16
		240	63.80	2.70	49.97	4.36	42.22	8.43	37.73	16.59	34.45	33.50	33.53	68.15	32.27	137.55
		300	65.26	3.54	51.92	5.75	44.34	11.48	39.94	23.36	37.32	48.47	36.51	99.63	36.17	203.08
		360	67.32	4.47	54.59	7.36	47.74	15.03	43.86	31.39	42.42	66.12	41.55	137.99	41.37	285.89
	Three	60	47.09	1.94	38.42	3.14	32.94	5.74	29.44	10.48	26.29	20.14	24.23	40.02	22.66	77.86
		120	37.12	2.68	27.56	4.31	21.58	8.03	18.24	14.83	15.79	28.66	14.80	57.11	13.98	111.97
		180	35.55	3.69	26.17	5.94	20.07	11.22	16.68	21.17	14.57	41.70	13.19	83.59	12.93	167.07
		240	37.82	4.94	27.72	7.99	21.21	15.48	17.39	30.01	15.75	60.37	14.57	123.32	14.43	249.99
		300	40.37	6.42	29.70	10.51	23.90	20.83	20.05	41.40	18.20	85.62	17.69	177.52	17.42	364.84
		360	42.98	8.13	32.83	13.38	26.54	26.97	23.13	54.90	22.07	115.34	21.43	241.75	21.35	501.83
	Four	60	30.20	3.04	22.58	4.90	18.21	8.97	15.09	16.26	13.30	30.82	11.74	61.19	10.88	118.80
		120	22.41	4.32	15.63	6.85	11.70	12.74	9.20	23.52	7.80	45.58	6.81	90.73	6.06	179.08
		180	22.53	5.90	15.53	9.43	11.49	17.82	8.92	33.66	7.29	66.47	6.60	133.70	6.12	269.71
		240	24.77	7.85	17.26	12.62	12.42	24.40	10.04	47.21	8.29	95.80	7.63	195.89	7.58	400.73
		300	26.39	10.20	19.36	16.57	14.32	32.61	11.48	64.92	10.09	134.33	9.65	280.15	9.29	574.56
		360	29.60	12.82	21.80	21.05	16.89	42.33	13.76	86.15	12.62	181.00	12.12	380.24	11.89	782.21

Table A.3 Simulation runs for San Francisco

Horizon			Simulation runs													
			1		2		4		8		16		32		64	
			Fail.	Run.	Fail.	Run.	Fail.	Run.	Fail.	Run.	Fail.	Run.	Fail.	Run.	Fail.	Run.
Vehicle fleet size	One	60	206.95	0.13	197.23	0.19	190.84	0.30	186.33	0.52	183.46	0.97	180.78	1.98	178.51	3.77
		120	187.94	0.17	175.53	0.25	168.94	0.44	162.86	0.81	159.03	1.63	156.45	3.47	154.09	6.92
		180	178.29	0.23	165.86	0.34	159.31	0.65	153.60	1.35	149.32	3.20	145.22	7.61	145.38	15.95
		240	172.44	0.31	161.05	0.48	153.79	0.96	149.79	2.12	145.51	5.47	143.86	14.99	142.84	33.53
		300	167.85	0.41	157.84	0.66	151.75	1.38	147.62	3.09	146.10	8.50	143.80	24.42	143.62	56.85
		360	167.17	0.54	158.23	0.89	152.31	1.90	149.46	4.33	146.50	11.85	145.27	34.56	146.00	81.58
	Two	60	132.59	0.26	116.68	0.40	104.71	0.70	96.75	1.25	90.54	2.38	85.96	4.85	84.79	9.21
		120	108.42	0.38	93.20	0.58	82.89	1.04	74.30	1.89	69.50	3.70	66.26	7.68	65.29	15.01
		180	97.18	0.52	83.91	0.78	73.98	1.47	67.48	2.84	62.54	5.94	61.38	12.90	60.10	25.79
		240	92.54	0.68	79.16	1.04	70.27	2.01	63.64	4.04	59.53	8.91	58.44	20.02	58.01	41.24
		300	91.07	0.87	77.99	1.34	68.47	2.66	62.89	5.48	59.60	12.48	58.28	28.13	58.83	58.40
		360	91.26	1.09	78.43	1.69	69.25	3.40	64.44	7.05	62.08	16.12	61.59	36.78	61.90	77.30
	Three	60	83.61	0.43	67.86	0.69	55.21	1.24	47.08	2.21	42.10	4.21	39.73	8.59	37.27	16.50
		120	65.00	0.67	50.77	1.03	40.22	1.88	33.60	3.42	29.56	6.71	28.25	13.86	27.75	27.20
		180	57.79	0.93	45.10	1.41	36.19	2.65	31.33	5.03	27.77	10.36	27.38	21.85	26.50	43.90
		240	54.36	1.21	42.83	1.84	34.35	3.55	30.04	7.02	28.21	14.87	27.44	31.73	27.58	64.54
		300	53.41	1.52	42.02	2.33	35.24	4.61	30.32	9.22	29.09	19.85	28.68	42.66	29.13	87.22
		360	55.10	1.87	42.69	2.88	35.87	5.78	31.92	11.73	30.39	25.33	30.82	54.43	31.31	112.18
	Four	60	55.46	0.65	40.22	1.04	30.84	1.88	25.33	3.36	22.16	6.45	20.37	13.12	18.98	25.22
		120	42.10	1.04	31.24	1.58	22.89	2.92	18.35	5.30	15.66	10.42	14.98	21.41	14.94	41.98
		180	39.09	1.44	27.84	2.17	21.16	4.11	17.27	7.72	15.77	15.67	15.06	32.74	14.44	65.37
		240	36.47	1.88	26.84	2.84	21.14	5.46	17.46	10.56	16.05	21.85	14.80	45.92	14.83	92.55
		300	36.18	2.34	27.23	3.56	21.61	6.96	18.34	13.60	16.41	28.52	15.31	60.29	14.77	122.43
		360	37.52	2.84	28.77	4.35	22.34	8.59	18.83	16.95	17.27	35.76	16.19	76.08	15.85	155.10

Table A.4 Dynamic lookahead horizons for DLA_{on} and DLA_{off}

		Minneapolis								San Francisco							
		DLA_{on}				DLA_{off}				DLA_{on}				DLA_{off}			
		1	2	3	4	1	2	3	4	1	2	3	4	1	2	3	4
Periods	0	60	240	120	120	360	300	240	300	180	180	300	60	180	60	120	300
	1	60	300	0	0	0	300	240	120	0	60	360	180	60	60	0	180
	2	300	360	360	0	0	120	0	0	360	300	180	300	360	0	120	60
	3	360	300	300	240	0	300	180	360	360	240	300	0	360	360	180	60
	4	240	240	240	180	300	300	240	360	300	360	240	180	300	360	360	360
	5	180	180	180	180	360	360	360	60	180	180	180	300	240	240	360	300
	6	120	120	180	120	300	180	360	300	120	120	180	120	240	360	360	300
	7	240	120	180	120	300	180	240	300	120	120	240	360	180	360	360	300
	8	180	60	120	240	300	180	180	360	120	60	180	120	120	360	360	360
	9	180	360	60	300	60	360	240	300	360	60	360	120	60	360	300	330
	10	60	60	240	300	300	360	360	360	240	300	360	60	240	300	360	360
	11	300	120	300	180	240	240	360	360	360	240	120	120	120	180	240	360
	12	300	240	240	180	300	240	300	360	240	360	360	180	180	240	60	300
	13	240	240	180	120	60	240	300	360	180	300	240	360	240	240	240	120
	14	180	180	300	180	120	60	180	300	360	180	180	180	300	180	180	180
	15	120	240	120	120	120	300	360	180	360	120	120	180	240	300	360	60
	16	120	180	120	120	180	240	300	360	120	60	60	60	60	240	240	120
	17	120	180	120	60	360	360	300	360	240	60	60	60	300	360	360	360
	18	120	180	120	60	60	60	360	360	120	120	120	120	120	240	360	240
	19	180	180	60	120	300	300	300	300	60	60	60	60	0	300	180	240
	20	180	120	240	60	240	120	180	180	60	60	240	120	180	120	240	240
	21	120	60	120	120	180	180	120	180	120	180	120	120	0	180	60	180
	22	60	120	120	120	120	120	120	120	120	120	120	120	0	60	60	120
	23	60	60	60	0	60	60	60	60	0	60	60	60	60	60	60	60

Table A.5 Dynamic lookahead horizons for manual DLAs

		Minneapolis								San Francisco								
		Ex Post DLA_{on}				Ex Post DLA_{off}				Ex Post DLA_{on}				Ex Post DLA_{off}				Apriori DLAs
		1	2	3	4	1	2	3	4	1	2	3	4	1	2	3	4	1–4
Periods	0	240	360	360	120	360	300	240	300	60	180	180	180	120	60	60	60	480
	1	240	300	240	120	180	360	180	360	60	60	60	60	300	300	300	300	420
	2	240	240	300	240	120	120	360	300	120	120	120	60	60	420	480	480	360
	3	240	240	120	60	120	300	60	60	60	60	60	60	60	60	60	60	300
	4	60	120	120	60	300	120	60	60	120	30	60	60	180	180	180	180	240
	5	180	180	120	120	240	240	240	240	120	60	60	60	480	360	300	300	180
	6	180	120	120	120	180	240	240	300	120	60	60	60	360	180	360	300	120
	7	240	240	180	180	300	240	300	480	120	120	60	60	300	300	300	420	60
	8	300	240	180	240	300	300	240	540	360	360	180	180	240	420	420	420	0
	9	300	240	180	240	300	240	360	420	360	360	300	240	360	360	420	420	180
	10	360	240	180	300	300	300	360	420	360	300	240	240	300	300	360	300	120
	11	300	240	180	240	300	300	300	360	300	240	180	180	240	300	300	300	60
	12	300	240	180	240	240	240	240	300	180	240	180	180	240	240	240	300	0
	13	240	180	180	180	180	180	180	180	180	180	120	120	180	240	240	240	240
	14	180	120	120	120	180	180	120	120	120	120	120	60	120	180	180	240	180
	15	120	120	120	120	60	120	120	120	60	60	60	60	60	120	120	180	120
	16	180	120	120	120	60	120	180	480	120	120	60	60	120	120	120	180	60
	17	240	240	180	180	180	300	420	420	360	360	120	360	360	420	420	420	0
	18	300	240	180	240	240	360	360	360	360	360	360	360	360	360	300	360	360
	19	300	300	180	180	180	300	300	300	300	300	300	180	300	300	300	300	300
	20	180	240	180	120	240	240	240	240	240	240	180	120	240	240	240	180	240
	21	180	180	180	180	180	180	180	180	180	180	120	120	180	180	180	180	180
	22	120	120	120	120	120	120	120	120	120	120	120	120	60	120	120	120	120
	23	60	60	60	60	60	60	60	60	60	60	60	60	60	60	60	60	60

Table A.6 Results of rollout algorithms in Minneapolis

Base policy		Simulation runs									
		1		2		4		8		16	
		Fail.	Run. (h)	Fail.	Run. (h)	Fail.	Run. (h)	Fail.	Run. (h)	Fail.	Run. (h)
STR(0.1)	60	254.02	0.01	249.56	0.03	249.33	0.06	237.17	0.13	222.55	0.32
	120	247.62	0.02	241.95	0.04	231.70	0.10	219.91	0.20	204.50	0.44
	180	247.22	0.03	238.47	0.05	232.83	0.11	222.91	0.23	210.79	0.52
	240	247.65	0.04	239.22	0.06	234.18	0.13	225.85	0.28	213.86	0.61
	300	245.85	0.04	238.67	0.07	237.38	0.16	223.27	0.32	214.87	0.71
	360	246.99	0.05	244.67	0.09	235.54	0.18	224.10	0.38	211.85	0.84
DLA_{off}		245.57	0.56	241.53	1.20	226.86	2.69	209.90	5.88	189.08	13.19

Table A.7 Results of rollout algorithms in San Francisco

Base policy		Simulation runs									
		1		2		4		8		16	
		Fail.	Run. (h)	Fail.	Run. (h)	Fail.	Run. (h)	Fail.	Run. (h)	Fail.	Run. (h)
STR(0.2)	60	301.90	0.00	285.88	0.00	271.40	0.00	257.80	0.01	240.80	0.01
	120	298.15	0.00	287.51	0.00	271.18	0.00	253.63	0.01	235.67	0.02
	180	291.19	0.00	286.02	0.00	270.24	0.01	251.02	0.01	229.48	0.02
	240	295.35	0.00	275.84	0.00	261.68	0.01	245.55	0.01	228.05	0.03
	300	295.42	0.00	281.01	0.00	261.09	0.01	239.96	0.02	229.54	0.04
	360	285.56	0.00	276.74	0.01	264.63	0.01	244.67	0.02	222.97	0.05
DLA_{off}		296.39	0.03	283.51	0.07	269.83	0.14	247.23	0.29	218.00	0.62
Base policy		Simulation runs									
		32		64		128		256			
		Fail.	Run. (h)	Fail.	Run. (h)	Fail.	Run. (h)	Fail.	Run. (h)		
STR(0.2)	60	218.81	0.03	210.86	0.06	199.22	0.12	194.66	0.24		
	120	216.81	0.04	204.84	0.08	194.32	0.16	185.88	0.32		
	180	217.63	0.05	198.38	0.10	187.09	0.21	184.59	0.40		
	240	214.09	0.06	195.90	0.13	181.96	0.26	175.06	0.52		
	300	207.58	0.08	191.65	0.16	178.04	0.33	177.72	0.67		
	360	210.89	0.09	191.14	0.20	178.00	0.40	165.16	0.83		
DLA_{off}		191.12	1.27	164.07	2.60	143.98	5.17	135.57	10.11		

Appendix B
Results

In this chapter, we present the detailed results of every experimental setup investigated. For every BSS, the results are separated by policy and vehicles fleet size. We distinguish STR, SLA_{on}, SLA_{off}, and DLA and one to four vehicles. Tables B.1 and B.2 provide an overview on the result tables.

In the result tables, we present various performance indicators. The average failed requests over all test instances are separated to failed rentals and failed returns. Additionally, we present the standard deviation and the standard error of the failed requests within the test instances. The failed request result in customer detours measured in minutes. Further, the average amounts of relocated bikes and served stations are shown. Last but not least, the maximum runtime observed over all decision points and test instances, and the average runtime over all test instances' MDPs are shown.

Table B.1 Results of STR and one vehicle in Minneapolis

Vehicles	STR	SLA_{on}	SLA_{off}	DLA
1	Table B.3	Table B.4	Table B.5	Table B.6
2	Tables B.7 and B.8	Tables B.9, B.10 and B.11	Tables B.12, B.13 and B.14	Table B.15
3	Tables B.16 and B.17	Tables B.18, B.19 and B.20	Tables B.21, B.22 and B.23	Table B.24
4	Tables B.25 and B.26	Tables B.27, B.28 and B.29	Tables B.30, B.31 and B.32	Table B.33

J. Brinkmann, *Active Balancing of Bike Sharing Systems*, Lecture Notes in Mobility,
https://doi.org/10.1007/978-3-030-35012-3

Table B.2 Results of STR and one vehicle in Minneapolis

Vehicles	STR	SLA_{on}	SLA_{off}	DLA
1	Table B.34	Table B.35	Table B.36	Table B.37
2	Tables B.38 and B.39	Tables B.40, B.41 and B.42	Tables B.43, B.44 and B.45	Table B.46
3	Tables B.47 and B.48	Tables B.49, B.50 and B.51	Tables B.52, B.53 and B.54	Table B.55
4	Tables B.56 and B.57	Tables B.58, B.59 and B.60	Tables B.61, B.62 and B.63	Table B.64

Table B.3 Results of STR and one vehicle in Minneapolis

	Do nothing	STR(0.1)	STR(0.2)	STR(0.3)	STR(0.4)	STR(0.5)
Failed requests	259.096	139.109	139.764	165.871	189.749	211.938
Failed rentals	137.838	78.586	79.299	91.305	106.247	120.811
Failed returns	121.258	60.523	60.465	74.566	83.502	91.127
Standard deviation	38.252	29.051	30.363	34.563	35.251	36.516
Standard error	1.210	0.919	0.960	1.093	1.115	1.155
Customer detours (min)	1107.354	610.364	609.979	715.985	840.561	957.726
Relocated bikes	–	190.071	279.220	385.419	447.159	485.078
Served stations	–	132.299	148.871	165.857	171.482	188.801
Max. runtime per k (s)	–	0.047	0.046	0.047	0.062	0.046
Avg. runtime per K (s)	–	0.135	0.137	0.140	0.143	0.149

Table B.4 Results of SLA_{on} and one vehicle in Minneapolis

	$SLA_{on}(60)$	$SLA_{on}(120)$	$SLA_{on}(180)$	$SLA_{on}(240)$	$SLA_{on}(300)$	$SLA_{on}(360)$
Failed requests	131.737	107.784	94.715	91.759	92.486	97.074
Failed rentals	71.619	58.438	51.598	50.196	50.737	53.479
Failed returns	60.118	49.346	43.117	41.563	41.749	43.595
Standard deviation	30.678	28.447	24.110	23.385	22.273	22.794
Standard error	0.970	0.900	0.762	0.740	0.704	0.721
Customer detours (min)	536.117	440.029	392.909	384.843	392.788	414.848
Relocated bikes	373.845	387.675	392.768	386.422	373.646	358.551
Served stations	44.272	46.776	58.314	80.793	110.793	143.200
Max. runtime per k (s)	1.170	0.655	0.725	0.870	1.900	1.389
Avg. runtime per K (s)	9.268	11.354	17.145	29.596	50.195	76.985

Table B.5 Results of SLA_{off} and one vehicle in Minneapolis

	$SLA_{off}(60)$	$SLA_{off}(120)$	$SLA_{off}(180)$	$SLA_{off}(240)$	$SLA_{off}(300)$	$SLA_{off}(360)$
Failed requests	123.866	108.693	99.969	98.327	100.117	103.872
Failed rentals	68.027	59.889	56.214	55.145	56.045	58.353
Failed returns	55.839	48.804	43.755	43.182	44.072	45.519
Standard deviation	27.911	26.407	24.855	24.050	24.058	24.150
Standard error	0.883	0.835	0.786	0.761	0.761	0.764
Customer detours (min)	498.906	437.348	408.776	403.850	412.133	431.650
Relocated bikes	193.769	204.462	213.588	219.292	219.984	217.735
Served stations	53.385	55.744	67.469	94.712	132.305	172.824
Max. runtime per k (s)	0.047	0.047	0.062	0.047	0.047	0.078
Avg. runtime per K (s)	0.150	0.156	0.168	0.190	0.222	0.265
	$SLA_{off}(420)$	$SLA_{off}(480)$	$SLA_{off}(540)$	$SLA_{off}(600)$	$SLA_{off}(660)$	$SLA_{off}(720)$
Failed requests	109.418	117.383	124.061	126.413	128.595	130.176
Failed rentals	61.894	67.017	70.989	72.354	73.530	74.624
Failed returns	47.524	50.366	53.072	54.059	55.065	55.552
Standard deviation	24.936	25.910	26.520	26.383	26.477	26.926
Standard error	0.789	0.819	0.839	0.834	0.837	0.851
Customer detours (min)	459.274	499.861	534.067	546.006	556.649	565.285
Relocated bikes	211.338	202.510	194.973	192.296	190.364	189.067
Served stations	211.588	250.800	288.235	293.642	294.366	295.125
Max. runtime per k (s)	0.078	0.062	0.078	0.066	0.078	0.063
Avg. runtime per K (s)	0.316	0.370	0.427	0.462	0.485	0.507

Table B.6 Results of the DLAs and one vehicle in Minneapolis

	A-priori$_{off}$	A-priori$_{on}$	Ex-post$_{off}$	Ex-post$_{on}$	DLA$_{off}$	DLA$_{on}$
Failed requests	114.043	104.613	102.293	89.637	92.032	86.446
Failed rentals	63.308	57.798	58.288	48.697	51.807	46.930
Failed returns	50.735	46.815	44.005	40.940	40.225	39.516
Standard deviation	26.125	25.809	25.030	23.291	23.708	23.374
Standard error	0.826	0.816	0.792	0.737	0.750	0.739
Customer detours (min)	473.179	445.850	414.815	376.060	369.266	363.760
Relocated bikes	215.876	390.845	216.077	393.406	239.616	389.502
Served stations	80.386	77.494	81.339	57.963	86.255	62.529
Max. runtime per k (s)	0.047	0.708	0.047	0.983	0.047	0.843
Avg. runtime per K (s)	0.168	16.413	0.183	20.278	0.174	17.357

Table B.7 Results of STR, two vehicles, and independent dispatching in Minneapolis

	STR(0.1)	STR(0.2)	STR(0.3)	STR(0.4)	STR(0.5)
Failed requests	93.971	86.796	113.192	148.572	186.410
Failed rentals	56.056	53.007	65.714	87.891	113.041
Failed returns	37.915	33.789	47.478	60.681	73.369
Standard deviation	24.470	26.735	33.913	35.023	37.587
Standard error	0.774	0.845	1.072	1.108	1.189
Customer detours (min)	411.124	379.437	484.229	666.818	876.788
Relocated bikes	290.496	445.337	672.561	817.002	930.234
Served stations	222.360	271.255	329.709	348.209	405.348
Max. runtime per k (s)	0.062	0.047	0.062	0.047	0.047
Avg. runtime per K (s)	0.139	0.141	0.149	0.157	0.173

Table B.8 Results of STR, two vehicles, and heuristic dispatching in Minneapolis

	STR(0.1)	STR(0.2)	STR(0.3)	STR(0.4)	STR(0.5)
Failed requests	88.825	79.773	94.920	140.015	174.842
Failed rentals	54.869	53.040	58.738	83.385	107.262
Failed returns	33.956	26.733	36.182	56.630	67.580
Standard deviation	23.791	24.231	28.018	33.595	35.812
Standard error	0.752	0.766	0.886	1.062	1.132
Customer detours (min)	397.605	416.937	630.111	824.606	824.606
Relocated bikes	264.088	640.873	801.691	901.724	901.724
Served stations	204.561	324.486	342.584	388.370	388.370
Max. runtime per k (s)	0.032	0.063	0.062	0.110	0.110
Avg. runtime per K (s)	0.150	0.180	0.192	0.220	0.220

Table B.9 Results of SLA_{on}, two vehicles, and independent dispatching in Minneapolis

	$SLA_{on}(60)$	$SLA_{on}(120)$	$SLA_{on}(180)$	$SLA_{on}(240)$	$SLA_{on}(300)$	$SLA_{on}(360)$
Failed requests	81.927	62.023	55.174	57.584	62.693	69.226
Failed rentals	47.161	35.423	31.574	32.874	36.143	40.003
Failed returns	34.766	26.600	23.600	24.710	26.550	29.223
Standard deviation	26.551	21.796	17.777	17.504	17.743	19.564
Standard error	0.840	0.689	0.562	0.554	0.561	0.619
Customer detours (min)	344.291	269.204	249.070	261.508	287.823	319.080
Relocated bikes	712.700	720.110	704.877	685.496	668.280	650.660
Served stations	94.302	104.046	138.783	195.434	253.546	315.083
Max. runtime per k (s)	0.484	0.952	0.827	1.125	2.871	3.432
Avg. runtime per K (s)	22.910	30.371	49.524	87.290	139.494	203.250

Table B.10 Results of SLA_{on}, two vehicles, and heuristic dispatching in Minneapolis

	$SLA_{on}(60)$	$SLA_{on}(120)$	$SLA_{on}(180)$	$SLA_{on}(240)$	$SLA_{on}(300)$	$SLA_{on}(360)$
Failed requests	57.342	38.231	33.804	33.527	36.508	41.555
Failed rentals	32.895	21.670	19.099	19.250	20.933	23.998
Failed returns	24.447	16.561	14.705	14.277	15.575	17.557
Standard deviation	20.500	15.532	12.966	11.902	11.959	12.762
Standard error	0.648	0.491	0.410	0.376	0.378	0.404
Customer detours (min)	239.987	167.779	154.429	159.203	175.989	201.911
Relocated bikes	703.320	708.534	698.848	675.383	642.807	610.780
Served stations	88.406	99.461	122.457	150.027	178.253	202.183
Max. runtime per k (s)	0.515	0.827	0.857	1.342	1.584	2.121
Avg. runtime per K (s)	22.721	30.477	45.443	68.146	99.627	137.991

Table B.11 Results of SLA_{on}, two vehicles, and optimal dispatching in Minneapolis

	$SLA_{on}(60)$	$SLA_{on}(120)$	$SLA_{on}(180)$	$SLA_{on}(240)$	$SLA_{on}(300)$	$SLA_{on}(360)$
Failed requests	63.701	42.104	37.210	38.825	42.198	48.131
Failed rentals	36.514	23.828	21.304	22.253	24.289	27.898
Failed returns	27.187	18.276	15.906	16.572	17.909	20.233
Standard deviation	23.294	17.032	14.215	14.095	14.769	15.845
Standard error	0.737	0.539	0.450	0.446	0.467	0.501
Customer detours (min)	265.230	182.659	169.451	181.088	198.440	227.531
Relocated bikes	717.927	719.054	702.606	672.133	636.910	598.854
Served stations	82.760	92.539	106.906	120.703	133.645	143.978
Max. runtime per k (s)	0.498	0.593	0.894	1.303	2.697	1.779
Avg. runtime per K (s)	24.709	33.644	50.088	74.275	106.482	145.346

Table B.12 Results of SLA_{off}, two vehicles, and independent dispatching in Minneapolis

	$SLA_{off}(60)$	$SLA_{off}(120)$	$SLA_{off}(180)$	$SLA_{off}(240)$	$SLA_{off}(300)$	$SLA_{off}(360)$
Failed requests	94.185	76.061	69.019	67.483	70.735	75.079
Failed rentals	53.354	43.661	39.660	38.141	39.335	42.104
Failed returns	40.831	32.400	29.359	29.342	31.400	32.975
Standard deviation	25.706	22.388	20.186	19.590	20.423	20.285
Standard error	0.813	0.708	0.638	0.619	0.646	0.641
Customer detours (min)	379.135	309.914	284.647	280.156	292.896	314.201
Relocated bikes	373.806	377.614	386.020	397.735	402.111	403.251
Served stations	99.491	104.197	125.519	176.556	247.061	320.569
Max. runtime per k (s)	0.078	0.062	0.078	0.063	0.063	0.093
Avg. runtime per K (s)	0.238	0.292	0.364	0.492	0.682	0.919
	$SLA_{off}(420)$	$SLA_{off}(480)$	$SLA_{off}(540)$	$SLA_{off}(600)$	$SLA_{off}(660)$	$SLA_{off}(720)$
Failed requests	81.917	90.414	99.753	104.770	107.891	110.700
Failed rentals	46.356	51.873	57.952	60.777	62.190	63.918
Failed returns	35.561	38.541	41.801	43.993	45.701	46.782
Standard deviation	21.423	23.011	23.709	24.182	24.372	24.994
Standard error	0.677	0.728	0.750	0.765	0.771	0.790
Customer detours (min)	348.033	389.013	434.486	457.428	471.637	484.671
Relocated bikes	393.170	375.253	365.629	364.130	362.896	360.557
Served stations	390.513	462.076	518.611	537.789	537.546	538.397
Max. runtime per k (s)	0.078	0.078	0.068	0.078	0.078	0.078
Avg. runtime per K (s)	1.193	1.510	1.806	2.026	2.143	2.255

Table B.13 Results of SLA_{off}, two vehicles, and heuristic dispatching in Minneapolis

	$SLA_{off}(60)$	$SLA_{off}(120)$	$SLA_{off}(180)$	$SLA_{off}(240)$	$SLA_{off}(300)$	$SLA_{off}(360)$
Failed requests	71.891	59.436	54.509	53.364	54.369	56.133
Failed rentals	44.364	37.119	34.249	33.271	33.711	34.845
Failed returns	27.527	22.317	20.260	20.093	20.658	21.288
Standard deviation	20.595	18.716	18.091	17.189	17.499	17.350
Standard error	0.651	0.592	0.572	0.544	0.553	0.549
Customer detours (min)	305.460	254.182	237.862	235.460	241.218	250.872
Relocated bikes	281.193	294.228	303.021	309.339	312.950	314.166
Served stations	81.617	90.883	113.001	141.284	168.455	196.233
Max. runtime per k (s)	0.063	0.063	0.063	0.063	0.063	0.063
Avg. runtime per K (s)	0.257	0.325	0.408	0.502	0.599	0.702
	$SLA_{off}(420)$	$SLA_{off}(480)$	$SLA_{off}(540)$	$SLA_{off}(600)$	$SLA_{off}(660)$	$SLA_{off}(720)$
Failed requests	59.213	64.581	70.596	75.493	77.935	80.694
Failed rentals	36.693	39.915	43.966	46.681	48.298	50.310
Failed returns	22.520	24.666	26.630	28.812	29.637	30.384
Standard deviation	17.762	18.860	19.409	20.447	20.885	20.855
Standard error	0.562	0.596	0.614	0.647	0.660	0.660
Customer detours (min)	266.301	291.592	322.222	344.957	357.946	373.190
Relocated bikes	312.316	305.785	299.383	296.219	295.551	292.746
Served stations	220.537	242.816	263.050	270.522	273.469	273.484
Max. runtime per k (s)	0.076	0.067	0.094	0.078	0.063	0.094
Avg. runtime per K (s)	0.798	0.899	1.009	1.104	1.172	1.227

Table B.14 Results of SLA_{off}, two vehicles, and optimal dispatching in Minneapolis

	$SLA_{off}(60)$	$SLA_{off}(120)$	$SLA_{off}(180)$	$SLA_{off}(240)$	$SLA_{off}(300)$	$SLA_{off}(360)$
Failed requests	70.965	59.607	57.096	58.417	61.915	66.109
Failed rentals	44.053	37.279	35.607	35.843	37.472	39.680
Failed returns	26.912	22.328	21.489	22.574	24.443	26.429
Standard deviation	19.772	18.878	18.199	17.335	17.786	18.438
Standard error	0.625	0.597	0.575	0.548	0.562	0.583
Customer detours (min)	301.704	254.723	247.927	253.786	269.957	289.187
Relocated bikes	284.004	295.936	299.911	300.364	298.807	295.249
Served stations	76.716	83.751	88.148	88.473	87.453	85.543
Max. runtime per k (s)	0.078	0.078	0.067	0.070	0.069	0.078
Avg. runtime per K (s)	0.331	0.424	0.518	0.611	0.699	0.789
	$SLA_{off}(420)$	$SLA_{off}(480)$	$SLA_{off}(540)$	$SLA_{off}(600)$	$SLA_{off}(660)$	$SLA_{off}(720)$
Failed requests	71.894	80.280	91.473	96.484	99.919	101.715
Failed rentals	43.094	47.953	54.672	57.949	60.073	61.063
Failed returns	28.800	32.327	36.801	38.535	39.846	40.652
Standard deviation	19.605	20.758	22.238	23.194	23.352	23.752
Standard error	0.620	0.656	0.703	0.733	0.738	0.751
Customer detours (min)	316.907	357.212	412.228	438.022	454.432	463.822
Relocated bikes	287.921	274.988	261.249	257.115	256.103	255.658
Served stations	82.656	78.126	71.513	68.465	67.252	66.026
Max. runtime per k (s)	0.073	0.080	0.078	0.078	0.094	0.078
Avg. runtime per K (s)	0.875	0.959	1.041	1.120	1.190	1.243

Table B.15 Results of the DLAs and two vehicles in Minneapolis

	A-priori$_{off}$	A-priori$_{on}$	Ex-post$_{off}$	Ex-post$_{on}$	DLA$_{off}$	DLA$_{on}$
Failed requests	84.217	43.585	54.570	34.413	52.842	30.738
Failed rentals	50.206	24.693	34.320	19.803	33.070	17.259
Failed returns	34.011	18.892	20.250	14.610	19.772	13.479
Standard deviation	22.125	15.729	18.622	13.625	17.523	12.237
Standard error	0.700	0.497	0.589	0.431	0.554	0.387
Customer detours (min)	356.142	192.392	235.389	153.558	230.055	144.430
Relocated bikes	273.711	678.233	310.262	707.316	313.274	670.983
Served stations	90.379	158.770	124.072	114.753	142.345	148.591
Max. runtime per k (s)	0.062	0.842	0.047	0.954	0.063	1.186
Avg. runtime per K (s)	0.376	41.238	0.500	41.554	0.496	52.980

Table B.16 Results of STR, three vehicles, and independent dispatching in Minneapolis

	STR(0.1)	STR(0.2)	STR(0.3)	STR(0.4)	STR(0.5)
Failed requests	77.033	67.012	84.544	125.577	169.459
Failed rentals	46.972	42.515	52.564	76.342	108.311
Failed returns	30.061	24.497	31.980	49.235	61.148
Standard deviation	23.789	27.175	34.000	43.125	40.339
Standard error	0.752	0.859	1.075	1.364	1.276
Customer detours (min)	332.895	292.150	370.287	569.934	826.139
Relocated bikes	364.182	565.077	903.459	1142.210	1359.695
Served stations	295.178	377.258	490.948	523.993	614.809
Max. runtime per k (s)	0.047	0.047	0.062	0.063	0.047
Avg. runtime per K (s)	0.142	0.145	0.158	0.170	0.199

Table B.17 Results of STR, three vehicles, and heuristic dispatching in Minneapolis

	STR(0.1)	STR(0.2)	STR(0.3)	STR(0.4)	STR(0.5)
Failed requests	67.855	52.773	58.550	90.108	146.733
Failed rentals	44.149	39.366	42.828	62.608	95.026
Failed returns	23.706	13.407	15.722	27.500	51.707
Standard deviation	20.710	20.921	21.717	28.620	32.464
Standard error	0.655	0.662	0.687	0.905	1.027
Customer detours (min)	305.650	274.413	428.955	711.824	711.824
Relocated bikes	292.928	767.623	1058.568	1265.608	1265.608
Served stations	245.668	452.319	520.113	587.555	587.555
Max. runtime per k (s)	0.047	0.062	0.062	0.063	0.063
Avg. runtime per K (s)	0.161	0.229	0.270	0.331	0.331

Table B.18 Results of SLA_{on}, three vehicles, and independent dispatching in Minneapolis

	$SLA_{on}(60)$	$SLA_{on}(120)$	$SLA_{on}(180)$	$SLA_{on}(240)$	$SLA_{on}(300)$	$SLA_{on}(360)$
Failed requests	55.544	41.237	38.750	43.145	48.579	55.457
Failed rentals	33.691	24.916	23.288	25.867	29.061	33.413
Failed returns	21.853	16.321	15.462	17.278	19.518	22.044
Standard deviation	20.662	15.879	13.653	14.383	15.403	16.485
Standard error	0.653	0.502	0.432	0.455	0.487	0.521
Customer detours (min)	246.460	193.093	187.362	208.792	235.373	269.815
Relocated bikes	1013.373	1004.080	973.694	947.047	933.721	922.115
Served stations	146.566	167.564	226.254	300.888	371.264	436.864
Max. runtime per k (s)	0.460	1.221	1.183	1.482	2.964	3.994
Avg. runtime per K (s)	39.783	56.721	95.794	157.564	232.785	319.207

Table B.19 Results of SLA_{on}, three vehicles, and heuristic dispatching in Minneapolis

	$SLA_{on}(60)$	$SLA_{on}(120)$	$SLA_{on}(180)$	$SLA_{on}(240)$	$SLA_{on}(300)$	$SLA_{on}(360)$
Failed requests	24.230	14.796	13.185	14.573	17.692	21.428
Failed rentals	14.583	8.808	7.878	8.895	11.049	13.151
Failed returns	9.647	5.988	5.307	5.678	6.643	8.277
Standard deviation	11.912	7.688	6.619	6.875	7.639	7.984
Standard error	0.377	0.243	0.209	0.217	0.242	0.252
Customer detours (min)	106.578	70.892	66.045	76.816	94.842	113.784
Relocated bikes	987.957	984.927	960.131	912.463	859.706	807.175
Served stations	132.471	152.736	183.866	220.404	256.456	286.300
Max. runtime per k (s)	0.453	1.419	1.887	2.560	3.037	2.617
Avg. runtime per K (s)	40.025	57.114	83.591	123.316	177.519	241.746

Table B.20 Results of SLA_{on}, three vehicles, and optimal dispatching in Minneapolis

	$SLA_{on}(60)$	$SLA_{on}(120)$	$SLA_{on}(180)$	$SLA_{on}(240)$	$SLA_{on}(300)$	$SLA_{on}(360)$
Failed requests	32.621	18.678	15.853	17.120	19.270	23.145
Failed rentals	19.690	11.274	9.590	10.452	11.830	14.291
Failed returns	12.931	7.404	6.263	6.668	7.440	8.854
Standard deviation	17.142	10.179	8.185	8.094	8.216	8.818
Standard error	0.542	0.322	0.259	0.256	0.260	0.279
Customer detours (min)	139.878	87.073	78.483	87.414	100.984	122.429
Relocated bikes	1009.905	1006.945	977.114	924.403	867.571	813.124
Served stations	118.677	135.658	155.531	176.070	195.299	208.972
Max. runtime per k (s)	0.602	0.771	1.014	1.388	2.044	2.418
Avg. runtime per K (s)	44.091	63.982	93.607	134.936	190.223	255.440

Table B.21 Results of SLA_{off}, three vehicles, and independent dispatching in Minneapolis

	$SLA_{off}(60)$	$SLA_{off}(120)$	$SLA_{off}(180)$	$SLA_{off}(240)$	$SLA_{off}(300)$	$SLA_{off}(360)$
Failed requests	86.766	67.563	61.034	59.151	61.819	65.847
Failed rentals	51.636	40.040	35.822	34.046	34.871	36.911
Failed returns	35.130	27.523	25.212	25.105	26.948	28.936
Standard deviation	26.595	21.271	19.470	17.805	17.462	18.315
Standard error	0.841	0.673	0.616	0.563	0.552	0.579
Customer detours (min)	355.874	281.502	256.818	252.248	263.414	281.609
Relocated bikes	907.471	810.024	747.852	701.030	666.611	636.209
Served stations	147.612	152.993	175.664	220.633	276.514	333.202
Max. runtime per k (s)	0.062	0.062	0.094	0.063	0.078	0.063
Avg. runtime per K (s)	0.453	0.591	0.729	0.897	1.093	1.311
	$SLA_{off}(420)$	$SLA_{off}(480)$	$SLA_{off}(540)$	$SLA_{off}(600)$	$SLA_{off}(660)$	$SLA_{off}(720)$
Failed requests	71.503	79.797	88.392	94.714	98.567	102.249
Failed rentals	39.942	45.263	51.161	54.425	56.406	58.586
Failed returns	31.561	34.534	37.231	40.289	42.161	43.663
Standard deviation	19.287	21.169	23.044	23.549	24.389	24.957
Standard error	0.610	0.669	0.729	0.745	0.771	0.789
Customer detours (min)	306.052	346.133	386.290	413.585	430.065	448.143
Relocated bikes	606.079	571.970	547.450	544.624	542.748	544.916
Served stations	391.712	483.078	551.081	569.334	576.252	573.742
Max. runtime per k (s)	0.078	0.094	0.094	0.093	0.094	0.078
Avg. runtime per K (s)	1.569	2.004	2.409	2.661	2.856	3.005

Table B.22 Results of SLA_{off}, three vehicles, and heuristic dispatching in Minneapolis

	$SLA_{off}(60)$	$SLA_{off}(120)$	$SLA_{off}(180)$	$SLA_{off}(240)$	$SLA_{off}(300)$	$SLA_{off}(360)$
Failed requests	53.889	44.715	40.841	38.677	38.231	38.479
Failed rentals	36.058	30.265	27.802	26.084	25.925	25.904
Failed returns	17.831	14.450	13.039	12.593	12.306	12.575
Standard deviation	17.268	16.765	15.542	14.503	13.964	13.985
Standard error	0.546	0.530	0.491	0.459	0.442	0.442
Customer detours (min)	240.337	200.807	187.017	179.047	179.425	181.264
Relocated bikes	313.976	326.833	337.490	345.492	351.698	356.466
Served stations	92.537	105.030	128.611	155.838	182.694	211.218
Max. runtime per k (s)	0.063	0.063	0.104	0.063	0.068	0.124
Avg. runtime per K (s)	0.404	0.573	0.748	0.926	1.095	1.267
	$SLA_{off}(420)$	$SLA_{off}(480)$	$SLA_{off}(540)$	$SLA_{off}(600)$	$SLA_{off}(660)$	$SLA_{off}(720)$
Failed requests	38.739	40.070	41.944	43.948	45.094	46.255
Failed rentals	26.160	27.229	28.507	30.042	30.591	31.639
Failed returns	12.579	12.841	13.437	13.906	14.503	14.616
Standard deviation	13.760	14.126	14.568	14.656	14.897	14.973
Standard error	0.435	0.447	0.461	0.463	0.471	0.473
Customer detours (min)	183.486	191.609	200.464	211.401	217.288	223.799
Relocated bikes	360.687	362.499	364.776	368.045	369.306	369.990
Served stations	237.792	263.899	289.257	297.872	305.380	308.506
Max. runtime per k (s)	0.078	0.078	0.093	0.069	0.072	0.069
Avg. runtime per K (s)	1.426	1.578	1.729	1.856	1.962	2.052

Table B.23 Results of SLA_{off}, three vehicles, and optimal dispatching in Minneapolis

	$SLA_{off}(60)$	$SLA_{off}(120)$	$SLA_{off}(180)$	$SLA_{off}(240)$	$SLA_{off}(300)$	$SLA_{off}(360)$
Failed requests	53.993	45.030	42.200	41.344	42.059	44.454
Failed rentals	35.964	30.150	28.313	27.418	27.886	29.299
Failed returns	18.029	14.880	13.887	13.926	14.173	15.155
Standard deviation	17.875	16.732	15.496	14.550	14.628	14.480
Standard error	0.565	0.529	0.490	0.460	0.463	0.458
Customer detours (min)	239.392	201.057	192.516	191.479	196.545	209.036
Relocated bikes	315.858	327.943	333.863	339.348	342.297	341.976
Served stations	85.068	94.012	101.074	106.365	109.492	110.014
Max. runtime per k (s)	0.080	0.078	0.078	0.046	0.078	0.078
Avg. runtime per K (s)	0.541	0.750	0.948	1.144	1.321	1.498
	$SLA_{off}(420)$	$SLA_{off}(480)$	$SLA_{off}(540)$	$SLA_{off}(600)$	$SLA_{off}(660)$	$SLA_{off}(720)$
Failed requests	47.281	52.044	60.225	66.027	70.119	72.835
Failed rentals	31.111	33.943	39.146	42.872	45.608	47.510
Failed returns	16.170	18.101	21.079	23.155	24.511	25.325
Standard deviation	14.867	15.892	17.093	18.073	18.838	19.049
Standard error	0.470	0.503	0.541	0.572	0.596	0.602
Customer detours (min)	224.388	246.897	287.360	315.957	337.405	352.834
Relocated bikes	340.460	334.806	325.395	321.780	320.131	318.704
Served stations	109.759	107.430	102.316	98.034	94.244	91.573
Max. runtime per k (s)	0.078	0.078	0.084	0.078	0.105	0.097
Avg. runtime per K (s)	1.668	1.821	1.954	2.096	2.235	2.347

Table B.24 Results of the DLAs and three vehicles in Minneapolis

	A-priori$_{off}$	A-priori$_{on}$	Ex-post$_{off}$	Ex-post$_{on}$	DLA$_{off}$	DLA$_{on}$
Failed requests	65.725	21.557	39.868	13.918	37.887	11.586
Failed rentals	41.883	12.679	27.060	8.228	25.671	6.762
Failed returns	23.842	8.878	12.808	5.690	12.216	4.824
Standard deviation	19.120	9.815	15.202	7.066	14.889	6.039
Standard error	0.605	0.310	0.481	0.223	0.471	0.191
Customer detours (min)	288.681	98.594	181.055	67.487	176.696	56.777
Relocated bikes	313.803	907.997	345.752	996.074	353.581	886.613
Served stations	98.487	228.432	138.458	161.813	164.368	233.630
Max. runtime per k (s)	0.063	1.061	0.047	0.827	0.078	1.877
Avg. runtime per K (s)	0.656	72.257	0.957	67.059	1.007	90.915

Table B.25 Results of STR, four vehicles, and independent dispatching in Minneapolis

	STR(0.1)	STR(0.2)	STR(0.3)	STR(0.4)	STR(0.5)
Failed requests	69.445	57.046	78.801	112.751	159.889
Failed rentals	43.053	36.728	50.069	69.636	104.325
Failed returns	26.392	20.318	28.732	43.115	55.564
Standard deviation	23.315	26.814	51.469	49.778	41.177
Standard error	0.737	0.848	1.628	1.574	1.302
Customer detours (min)	295.438	247.906	353.650	515.936	790.407
Relocated bikes	433.029	672.094	1111.740	1424.190	1780.787
Served stations	363.059	483.200	634.953	690.984	817.161
Max. runtime per k (s)	0.046	0.047	0.047	0.047	0.047
Avg. runtime per K (s)	0.144	0.150	0.166	0.184	0.228

Table B.26 Results of STR, four vehicles, and heuristic dispatching in Minneapolis

	STR(0.1)	STR(0.2)	STR(0.3)	STR(0.4)	STR(0.5)
Failed requests	58.726	40.677	40.168	57.501	122.655
Failed rentals	39.651	32.391	33.025	48.086	85.918
Failed returns	19.075	8.286	7.143	9.415	36.737
Standard deviation	19.642	18.574	18.253	21.012	30.328
Standard error	0.621	0.587	0.577	0.664	0.959
Customer detours (min)	265.729	197.636	299.017	621.160	621.160
Relocated bikes	305.480	830.924	1202.466	1570.640	1570.640
Served stations	270.066	549.638	687.969	797.498	797.498
Max. runtime per k (s)	0.062	0.047	0.063	0.062	0.062
Avg. runtime per K (s)	0.173	0.279	0.376	0.497	0.497

Table B.27 Results of SLA_{on}, four vehicles, and independent dispatching in Minneapolis

	$SLA_{on}(60)$	$SLA_{on}(120)$	$SLA_{on}(180)$	$SLA_{on}(240)$	$SLA_{on}(300)$	$SLA_{on}(360)$
Failed requests	41.088	31.173	31.289	35.020	40.762	48.212
Failed rentals	26.367	19.964	19.650	22.102	25.378	29.832
Failed returns	14.721	11.209	11.639	12.918	15.384	18.380
Standard deviation	15.872	13.517	11.595	12.342	12.881	14.117
Standard error	0.502	0.427	0.367	0.390	0.407	0.446
Customer detours (min)	191.750	154.301	158.033	178.806	205.699	241.298
Relocated bikes	1282.701	1260.299	1221.321	1196.870	1174.436	1159.745
Served stations	199.547	231.021	311.783	392.192	476.758	556.508
Max. runtime per k (s)	0.920	0.874	1.045	2.668	3.479	3.697
Avg. runtime per K (s)	59.205	89.037	151.968	235.484	339.193	460.964

Table B.28 Results of SLA_{on}, four vehicles, and heuristic dispatching in Minneapolis

	$SLA_{on}(60)$	$SLA_{on}(120)$	$SLA_{on}(180)$	$SLA_{on}(240)$	$SLA_{on}(300)$	$SLA_{on}(360)$
Failed requests	11.744	6.812	6.596	7.627	9.647	12.117
Failed rentals	7.445	4.265	4.166	4.944	6.329	8.049
Failed returns	4.299	2.547	2.430	2.683	3.318	4.068
Standard deviation	6.928	4.442	4.197	4.464	4.806	5.803
Standard error	0.219	0.140	0.133	0.141	0.152	0.184
Customer detours (min)	53.815	34.589	35.180	43.046	55.125	69.440
Relocated bikes	1252.459	1241.583	1202.754	1138.012	1062.548	996.913
Served stations	175.475	206.129	247.145	295.258	343.757	385.335
Max. runtime per k (s)	0.499	2.215	2.215	2.136	3.307	3.120
Avg. runtime per K (s)	61.193	90.728	133.701	195.889	280.155	380.242

Table B.29 Results of SLA_{on}, four vehicles, and optimal dispatching in Minneapolis

	$SLA_{on}(60)$	$SLA_{on}(120)$	$SLA_{on}(180)$	$SLA_{on}(240)$	$SLA_{on}(300)$	$SLA_{on}(360)$
Failed requests	19.347	9.764	7.941	8.855	10.675	12.795
Failed rentals	12.316	6.107	5.066	5.734	6.975	8.455
Failed returns	7.031	3.657	2.875	3.121	3.700	4.340
Standard deviation	12.858	7.170	5.357	5.244	5.598	5.892
Standard error	0.407	0.227	0.169	0.166	0.177	0.186
Customer detours (min)	84.083	46.981	40.306	47.839	58.738	72.463
Relocated bikes	1272.835	1261.674	1218.752	1149.828	1079.029	1013.474
Served stations	151.976	179.616	206.928	235.004	259.525	279.050
Max. runtime per k (s)	0.581	1.143	1.208	1.656	2.091	2.597
Avg. runtime per K (s)	67.268	101.980	149.862	216.412	303.207	406.423

Table B.30 Results of SLA_{off}, four vehicles, and independent dispatching in Minneapolis

	$SLA_{off}(60)$	$SLA_{off}(120)$	$SLA_{off}(180)$	$SLA_{off}(240)$	$SLA_{off}(300)$	$SLA_{off}(360)$
Failed requests	84.291	64.725	57.436	56.891	59.289	62.838
Failed rentals	49.722	38.015	33.420	32.544	33.152	34.818
Failed returns	34.569	26.710	24.016	24.347	26.137	28.020
Standard deviation	26.566	20.779	17.604	17.268	17.754	18.349
Standard error	0.840	0.657	0.557	0.546	0.561	0.580
Customer detours (min)	350.379	273.895	246.639	246.598	256.270	272.230
Relocated bikes	696.252	652.995	655.614	679.223	712.920	731.681
Served stations	198.350	204.101	231.684	281.141	334.559	392.171
Max. runtime per k (s)	0.063	0.063	0.072	0.064	0.069	0.069
Avg. runtime per K (s)	0.500	0.693	0.926	1.207	1.509	1.853
	$SLA_{off}(420)$	$SLA_{off}(480)$	$SLA_{off}(540)$	$SLA_{off}(600)$	$SLA_{off}(660)$	$SLA_{off}(720)$
Failed requests	67.485	74.137	82.777	87.883	92.433	95.543
Failed rentals	37.418	42.279	48.005	50.956	53.471	55.285
Failed returns	30.067	31.858	34.772	36.927	38.962	40.258
Standard deviation	19.173	20.279	21.821	22.419	23.415	23.345
Standard error	0.606	0.641	0.690	0.709	0.740	0.738
Customer detours (min)	291.563	324.412	363.456	387.871	407.721	421.380
Relocated bikes	725.667	683.871	641.940	643.870	638.991	640.760
Served stations	455.178	558.433	653.958	679.624	696.176	696.212
Max. runtime per k (s)	0.078	0.078	0.107	0.094	0.094	0.078
Avg. runtime per K (s)	2.250	2.858	3.515	3.917	4.263	4.492

Table B.31 Results of SLA_{off}, four vehicles, and heuristic dispatching in Minneapolis

	$SLA_{off}(60)$	$SLA_{off}(120)$	$SLA_{off}(180)$	$SLA_{off}(240)$	$SLA_{off}(300)$	$SLA_{off}(360)$
Failed requests	48.427	40.286	36.401	34.238	32.703	32.245
Failed rentals	34.482	28.846	26.186	24.386	23.270	22.977
Failed returns	13.945	11.440	10.215	9.852	9.433	9.268
Standard deviation	16.462	15.810	14.648	13.899	13.422	12.969
Standard error	0.521	0.500	0.463	0.440	0.424	0.410
Customer detours (min)	226.071	188.668	172.696	164.649	158.403	157.575
Relocated bikes	329.806	342.193	352.965	360.730	368.606	375.412
Served stations	95.646	109.556	130.202	152.693	173.619	195.981
Max. runtime per k (s)	0.063	0.063	0.063	0.078	0.093	0.078
Avg. runtime per K (s)	0.602	0.908	1.211	1.503	1.782	2.054
	$SLA_{off}(420)$	$SLA_{off}(480)$	$SLA_{off}(540)$	$SLA_{off}(600)$	$SLA_{off}(660)$	$SLA_{off}(720)$
Failed requests	31.677	31.939	32.268	32.949	33.780	33.979
Failed rentals	22.655	23.061	23.372	23.725	24.368	24.464
Failed returns	9.022	8.878	8.896	9.224	9.412	9.515
Standard deviation	12.611	12.835	12.701	12.480	12.635	12.819
Standard error	0.399	0.406	0.402	0.395	0.400	0.405
Customer detours (min)	154.807	158.076	160.412	163.315	167.191	167.968
Relocated bikes	381.130	385.626	391.517	397.422	400.851	401.730
Served stations	211.961	229.789	251.932	263.820	273.863	278.241
Max. runtime per k (s)	0.071	0.094	0.094	0.262	0.078	0.093
Avg. runtime per K (s)	2.293	2.528	2.771	2.958	3.135	3.291

Table B.32 Results of SLA_{off}, four vehicles, and optimal dispatching in Minneapolis

	$SLA_{off}(60)$	$SLA_{off}(120)$	$SLA_{off}(180)$	$SLA_{off}(240)$	$SLA_{off}(300)$	$SLA_{off}(360)$
Failed requests	48.255	40.371	36.881	35.159	34.859	35.266
Failed rentals	33.761	28.332	26.035	24.762	24.318	24.679
Failed returns	14.494	12.039	10.846	10.397	10.541	10.587
Standard deviation	16.726	15.698	14.783	13.734	13.221	12.879
Standard error	0.529	0.496	0.467	0.434	0.418	0.407
Customer detours (min)	220.713	186.288	174.117	169.668	169.251	172.495
Relocated bikes	330.742	342.354	350.906	357.587	363.869	366.745
Served stations	87.242	96.860	105.879	113.221	118.748	121.442
Max. runtime per k (s)	0.078	0.081	0.078	0.109	0.078	0.094
Avg. runtime per K (s)	0.813	1.176	1.523	1.852	2.161	2.457
	$SLA_{off}(420)$	$SLA_{off}(480)$	$SLA_{off}(540)$	$SLA_{off}(600)$	$SLA_{off}(660)$	$SLA_{off}(720)$
Failed requests	36.890	38.165	41.143	43.880	46.123	47.871
Failed rentals	25.891	26.892	29.202	31.070	32.746	34.022
Failed returns	10.999	11.273	11.941	12.810	13.377	13.849
Standard deviation	13.279	12.858	13.764	13.959	14.305	14.777
Standard error	0.420	0.407	0.435	0.441	0.452	0.467
Customer detours (min)	183.737	190.343	207.229	221.655	234.045	242.214
Relocated bikes	369.360	370.874	369.570	370.182	372.361	372.269
Served stations	121.576	123.078	123.666	121.642	119.877	117.679
Max. runtime per k (s)	0.109	0.078	0.094	0.081	0.227	0.095
Avg. runtime per K (s)	2.739	2.994	3.213	3.429	3.627	3.804

Table B.33 Results of the DLAs and four vehicles in Minneapolis

	A-priori$_{off}$	A-priori$_{on}$	Ex-post$_{off}$	Ex-post$_{on}$	DLA$_{off}$	DLA$_{on}$
Failed requests	58.086	12.471	32.655	6.710	32.391	6.083
Failed rentals	39.112	7.693	22.866	4.269	23.048	3.797
Failed returns	18.974	4.778	9.789	2.441	9.343	2.286
Standard deviation	17.635	6.697	13.249	4.359	13.198	4.160
Standard error	0.558	0.212	0.419	0.138	0.417	0.132
Customer detours (min)	264.528	58.219	153.979	34.960	156.861	30.454
Relocated bikes	335.163	1115.653	369.863	1238.932	374.785	1103.192
Served stations	102.476	292.461	147.120	211.788	182.251	298.094
Max. runtime per k (s)	0.069	1.202	0.062	1.083	0.078	2.172
Avg. runtime per K (s)	1.055	110.320	1.834	119.148	1.774	113.876

Table B.34 Results of STR and one vehicle in San Francisco

	Do nothing	STR(0.1)	STR(0.2)	STR(0.3)	STR(0.4)	STR(0.5)
Failed requests	342.852	218.535	210.956	220.081	238.437	257.626
Failed rentals	147.765	97.887	92.894	93.153	99.010	110.932
Failed returns	195.087	120.648	118.062	126.928	139.427	146.694
Standard deviation	44.685	35.668	35.696	35.669	37.265	38.146
Standard error	1.413	1.128	1.129	1.128	1.178	1.206
Customer detours (min)	1215.721	781.408	761.776	794.510	862.607	940.743
Relocated bikes	–	172.659	247.512	347.054	420.552	480.540
Served stations	–	120.383	129.529	141.177	151.225	175.057
Max. runtime per k (s)	–	0.016	0.031	0.047	0.032	0.047
Avg. runtime per K (s)	–	0.043	0.043	0.044	0.044	0.046

Table B.35 Results of SLA_{on} and one vehicle in San Francisco

	$SLA_{on}(60)$	$SLA_{on}(120)$	$SLA_{on}(180)$	$SLA_{on}(240)$	$SLA_{on}(300)$	$SLA_{on}(360)$
Failed requests	180.775	156.454	145.216	143.859	143.796	145.270
Failed rentals	83.442	71.132	65.647	64.903	65.000	65.583
Failed returns	97.333	85.322	79.569	78.956	78.796	79.687
Standard deviation	36.492	33.229	32.009	31.819	31.991	32.276
Standard error	1.154	1.051	1.012	1.006	1.012	1.021
Customer detours (min)	602.310	509.593	470.880	468.672	469.972	477.312
Relocated bikes	417.938	434.036	432.850	422.616	407.562	388.942
Served stations	50.423	68.851	110.020	170.543	235.173	302.640
Max. runtime per k (s)	0.125	0.352	0.203	0.214	0.245	0.270
Avg. runtime per K (s)	1.976	3.474	7.612	14.989	24.420	34.556

Table B.36 Results of SLA_{off} and one vehicle in San Francisco

	$SLA_{off}(60)$	$SLA_{off}(120)$	$SLA_{off}(180)$	$SLA_{off}(240)$	$SLA_{off}(300)$	$SLA_{off}(360)$
Failed requests	182.013	158.895	143.213	135.102	133.154	134.716
Failed rentals	88.166	75.769	66.898	62.391	60.579	60.961
Failed returns	93.847	83.126	76.315	72.711	72.575	73.755
Standard deviation	36.459	34.427	32.968	32.015	31.571	32.185
Standard error	1.153	1.089	1.043	1.012	0.998	1.018
Customer detours (min)	621.772	532.709	474.575	443.002	434.563	438.776
Relocated bikes	198.114	219.499	234.014	241.565	243.106	240.607
Served stations	51.996	57.403	68.076	82.307	97.750	111.726
Max. runtime per k (s)	0.016	0.032	0.047	0.047	0.047	0.047
Avg. runtime per K (s)	0.049	0.051	0.054	0.058	0.063	0.068
	$SLA_{off}(420)$	$SLA_{off}(480)$	$SLA_{off}(540)$	$SLA_{off}(600)$	$SLA_{off}(660)$	$SLA_{off}(720)$
Failed requests	138.308	140.520	140.235	142.150	143.586	144.044
Failed rentals	62.712	63.414	63.273	64.577	65.488	66.027
Failed returns	75.596	77.106	76.962	77.573	78.098	78.017
Standard deviation	32.269	32.433	32.638	32.727	32.984	33.656
Standard error	1.020	1.026	1.032	1.035	1.043	1.064
Customer detours (min)	453.108	460.464	458.447	464.950	471.106	474.714
Relocated bikes	236.513	235.245	236.226	236.026	234.348	234.499
Served stations	125.339	133.775	134.732	136.843	139.813	142.337
Max. runtime per k (s)	0.047	0.047	0.047	0.047	0.047	0.062
Avg. runtime per K (s)	0.072	0.077	0.081	0.084	0.087	0.090

Table B.37 Results of the DLAs and one vehicle in San Francisco

	A-priori$_{\text{off}}$	A-priori$_{\text{on}}$	Ex-post$_{\text{off}}$	Ex-post$_{\text{on}}$	DLA$_{\text{off}}$	DLA$_{\text{on}}$
Failed requests	160.506	148.208	146.335	159.143	123.750	133.285
Failed rentals	74.573	62.947	65.803	71.643	55.299	57.837
Failed returns	85.933	85.261	80.532	87.500	68.451	75.448
Standard deviation	31.940	30.580	32.921	33.721	31.107	30.362
Standard error	1.010	0.967	1.041	1.066	0.984	0.960
Customer detours (min)	568.962	493.700	483.401	523.225	408.920	427.258
Relocated bikes	228.587	420.432	237.109	415.593	272.985	413.164
Served stations	85.434	166.046	59.360	57.076	108.629	154.573
Max. runtime per k (s)	0.046	0.172	0.033	0.250	0.063	0.249
Avg. runtime per K (s)	0.055	8.146	0.057	3.187	0.055	12.961

Table B.38 Results of STR, two vehicles, and independent dispatching in San Francisco

	STR(0.1)	STR(0.2)	STR(0.3)	STR(0.4)	STR(0.5)
Failed requests	165.735	147.950	148.703	167.264	235.117
Failed rentals	80.889	73.600	72.969	82.722	107.730
Failed returns	84.846	74.350	75.734	84.542	127.387
Standard deviation	30.176	29.458	30.553	34.050	47.000
Standard error	0.954	0.932	0.966	1.077	1.486
Customer detours (min)	601.895	555.622	568.630	652.456	891.717
Relocated bikes	248.356	352.146	511.342	684.946	962.044
Served stations	190.971	213.778	261.153	316.973	416.627
Max. runtime per k (s)	0.062	0.063	0.016	0.033	0.047
Avg. runtime per K (s)	0.045	0.045	0.045	0.047	0.054

Table B.39 Results of STR, two vehicles, and independent dispatching in San Francisco

	STR(0.1)	STR(0.2)	STR(0.3)	STR(0.4)	STR(0.5)
Failed requests	168.196	147.675	147.774	162.283	203.399
Failed rentals	82.035	74.377	73.170	80.664	95.246
Failed returns	86.161	73.298	74.604	81.619	108.153
Standard deviation	29.216	28.858	29.362	31.166	35.200
Standard error	0.924	0.913	0.929	0.986	1.113
Customer detours (min)	607.805	564.249	631.298	786.551	786.551
Relocated bikes	224.597	459.548	617.244	789.614	789.614
Served stations	170.881	241.466	290.305	361.066	361.066
Max. runtime per k (s)	0.047	0.047	0.031	0.047	0.047
Avg. runtime per K (s)	0.049	0.049	0.051	0.058	0.058

Table B.40 Results of SLA_{on}, two vehicles, and independent dispatching in San Francisco

	$SLA_{on}(60)$	$SLA_{on}(120)$	$SLA_{on}(180)$	$SLA_{on}(240)$	$SLA_{on}(300)$	$SLA_{on}(360)$
Failed requests	102.616	85.196	81.413	82.745	84.479	86.866
Failed rentals	58.259	48.186	46.363	47.916	48.757	50.238
Failed returns	44.357	37.010	35.050	34.829	35.722	36.628
Standard deviation	30.874	27.204	26.774	26.451	26.962	26.004
Standard error	0.976	0.860	0.847	0.836	0.853	0.822
Customer detours (min)	356.533	295.725	287.068	294.426	301.724	311.964
Relocated bikes	792.680	826.821	838.005	828.537	807.579	786.342
Served stations	110.645	156.183	231.834	331.819	433.279	524.926
Max. runtime per k (s)	0.125	0.165	0.218	0.328	1.217	0.889
Avg. runtime per K (s)	4.968	9.348	18.722	33.259	49.831	66.161

Table B.41 Results of SLA_{on}, two vehicles, and heuristic dispatching in San Francisco

	$SLA_{on}(60)$	$SLA_{on}(120)$	$SLA_{on}(180)$	$SLA_{on}(240)$	$SLA_{on}(300)$	$SLA_{on}(360)$
Failed requests	85.961	66.264	61.383	58.444	58.283	61.593
Failed rentals	44.028	31.990	30.331	29.676	30.614	32.272
Failed returns	41.933	34.274	31.052	28.768	27.669	29.321
Standard deviation	25.608	21.923	21.051	20.054	20.116	19.907
Standard error	0.810	0.693	0.666	0.634	0.636	0.629
Customer detours (min)	279.453	208.767	199.575	192.468	195.331	208.528
Relocated bikes	762.459	778.672	783.031	771.937	751.004	723.812
Served stations	99.363	128.086	164.519	204.985	246.755	288.206
Max. runtime per k (s)	0.141	0.187	0.203	0.312	0.967	1.076
Avg. runtime per K (s)	4.853	7.675	12.905	20.016	28.132	36.783

Table B.42 Results of SLA_{on}, two vehicles, and optimal dispatching in San Francisco

	$SLA_{on}(60)$	$SLA_{on}(120)$	$SLA_{on}(180)$	$SLA_{on}(240)$	$SLA_{on}(300)$	$SLA_{on}(360)$
Failed requests	96.686	77.212	72.809	70.833	72.447	75.909
Failed rentals	47.861	36.112	33.488	33.964	35.276	37.888
Failed returns	48.825	41.100	39.321	36.869	37.171	38.021
Standard deviation	29.205	25.880	25.023	24.680	24.761	25.791
Standard error	0.924	0.818	0.791	0.780	0.783	0.816
Customer detours (min)	310.697	239.110	226.753	226.298	234.337	249.247
Relocated bikes	758.219	763.107	757.911	739.014	714.831	678.781
Served stations	89.510	108.115	126.194	145.688	165.496	184.825
Max. runtime per k (s)	0.149	0.296	0.239	0.250	0.281	0.312
Avg. runtime per K (s)	5.037	8.040	12.683	18.776	25.568	32.711

Table B.43 Results of SLA_{off}, two vehicles, and independent dispatching in San Francisco

	$SLA_{off}(60)$	$SLA_{off}(120)$	$SLA_{off}(180)$	$SLA_{off}(240)$	$SLA_{off}(300)$	$SLA_{off}(360)$
Failed requests	121.067	90.713	72.402	64.880	63.619	66.369
Failed rentals	69.576	51.419	41.617	37.031	36.129	37.827
Failed returns	51.491	39.294	30.785	27.849	27.490	28.542
Standard deviation	32.578	27.122	23.873	21.080	21.034	21.808
Standard error	1.030	0.858	0.755	0.667	0.665	0.690
Customer detours (min)	424.769	311.301	255.691	230.598	223.891	234.101
Relocated bikes	378.213	406.474	421.147	433.441	434.595	426.164
Served stations	93.285	105.064	124.657	155.557	187.227	221.321
Max. runtime per k (s)	0.063	0.062	0.047	0.063	0.063	0.063
Avg. runtime per K (s)	0.075	0.095	0.117	0.141	0.168	0.198
	$SLA_{off}(420)$	$SLA_{off}(480)$	$SLA_{off}(540)$	$SLA_{off}(600)$	$SLA_{off}(660)$	$SLA_{off}(720)$
Failed requests	71.213	74.539	78.132	80.643	82.346	84.171
Failed rentals	40.704	42.491	44.043	46.034	47.316	48.617
Failed returns	30.509	32.048	34.089	34.609	35.030	35.554
Standard deviation	22.775	21.900	21.995	22.134	23.161	23.232
Standard error	0.720	0.693	0.696	0.700	0.732	0.735
Customer detours (min)	252.028	266.048	281.115	290.806	296.983	304.914
Relocated bikes	420.751	417.146	426.866	426.085	428.010	428.142
Served stations	248.842	265.752	265.184	266.012	267.791	273.574
Max. runtime per k (s)	0.063	0.047	0.062	0.047	0.062	0.047
Avg. runtime per K (s)	0.229	0.257	0.276	0.294	0.309	0.327

Table B.44 Results of SLA_{off}, two vehicles, and heuristic dispatching in San Francisco

	$SLA_{off}(60)$	$SLA_{off}(120)$	$SLA_{off}(180)$	$SLA_{off}(240)$	$SLA_{off}(300)$	$SLA_{off}(360)$
Failed requests	106.876	78.172	63.176	58.302	56.857	56.785
Failed rentals	62.388	44.993	35.806	32.787	31.463	31.573
Failed returns	44.488	33.179	27.370	25.515	25.394	25.212
Standard deviation	28.909	25.051	21.666	20.761	20.157	19.983
Standard error	0.914	0.792	0.685	0.657	0.637	0.632
Customer detours (min)	373.395	265.247	210.974	194.104	186.776	187.830
Relocated bikes	278.600	306.564	326.087	335.284	339.064	339.202
Served stations	72.918	85.733	104.745	125.742	146.682	167.285
Max. runtime per k (s)	0.062	0.062	0.062	0.047	0.062	0.078
Avg. runtime per K (s)	0.076	0.099	0.121	0.144	0.166	0.188
	$SLA_{off}(420)$	$SLA_{off}(480)$	$SLA_{off}(540)$	$SLA_{off}(600)$	$SLA_{off}(660)$	$SLA_{off}(720)$
Failed requests	58.323	60.909	63.827	65.983	66.943	66.986
Failed rentals	32.825	35.253	37.946	40.217	41.411	41.850
Failed returns	25.498	25.656	25.881	25.766	25.532	25.136
Standard deviation	20.188	20.520	20.915	21.127	21.837	21.504
Standard error	0.638	0.649	0.661	0.668	0.691	0.680
Customer detours (min)	193.612	207.286	224.441	233.902	238.282	240.039
Relocated bikes	339.432	337.649	336.157	336.889	338.426	338.171
Served stations	184.042	196.456	197.770	195.065	197.068	198.463
Max. runtime per k (s)	0.063	0.063	0.047	0.078	0.063	0.062
Avg. runtime per K (s)	0.211	0.229	0.245	0.257	0.271	0.284

Table B.45 Results of SLA_{off}, two vehicles, and optimal dispatching in San Francisco

	$SLA_{off}(60)$	$SLA_{off}(120)$	$SLA_{off}(180)$	$SLA_{off}(240)$	$SLA_{off}(300)$	$SLA_{off}(360)$
Failed requests	105.864	79.199	66.680	63.042	62.321	64.460
Failed rentals	63.399	47.029	39.177	36.835	36.040	37.325
Failed returns	42.465	32.170	27.503	26.207	26.281	27.135
Standard deviation	29.503	26.220	24.845	23.566	22.957	23.550
Standard error	0.933	0.829	0.786	0.745	0.726	0.745
Customer detours (min)	380.341	281.385	234.230	219.741	216.282	223.146
Relocated bikes	277.499	304.496	319.297	325.059	325.196	321.848
Served stations	67.002	75.445	83.128	86.692	87.810	87.757
Max. runtime per k (s)	0.017	0.020	0.034	0.031	0.016	0.016
Avg. runtime per K (s)	0.082	0.108	0.134	0.157	0.178	0.197
	$SLA_{off}(420)$	$SLA_{off}(480)$	$SLA_{off}(540)$	$SLA_{off}(600)$	$SLA_{off}(660)$	$SLA_{off}(720)$
Failed requests	68.410	75.789	85.689	94.064	97.412	97.327
Failed rentals	40.040	45.346	53.333	60.504	63.220	64.072
Failed returns	28.370	30.443	32.356	33.560	34.192	33.255
Standard deviation	23.460	23.317	25.747	26.358	27.893	27.343
Standard error	0.742	0.737	0.814	0.834	0.882	0.865
Customer detours (min)	239.287	276.626	324.954	362.907	377.244	377.601
Relocated bikes	315.758	307.127	298.119	292.703	289.673	289.942
Served stations	84.880	82.507	74.405	72.190	72.399	72.395
Max. runtime per k (s)	0.016	0.016	0.031	0.017	0.047	0.017
Avg. runtime per K (s)	0.214	0.230	0.244	0.257	0.270	0.281

Table B.46 Results of the DLAs and two vehicles in San Francisco

	A-priori$_{off}$	A-priori$_{on}$	Ex-post$_{off}$	Ex-post$_{on}$	DLA$_{off}$	DLA$_{on}$
Failed requests	131.155	77.470	57.989	84.122	54.396	50.017
Failed rentals	69.647	37.033	30.611	44.753	29.752	24.189
Failed returns	61.508	40.437	27.378	39.369	24.644	25.828
Standard deviation	29.594	22.243	20.714	25.477	20.222	19.094
Standard error	0.936	0.703	0.655	0.806	0.639	0.604
Customer detours (min)	483.900	261.005	188.849	276.073	177.871	157.304
Relocated bikes	264.191	753.699	342.296	756.903	344.574	767.683
Served stations	83.744	241.548	111.329	111.488	128.850	210.635
Max. runtime per k (s)	0.047	0.187	0.033	0.218	0.062	0.442
Avg. runtime per K (s)	0.116	13.456	0.147	7.855	0.146	16.216

Table B.47 Results of STR, three vehicles, and independent dispatching in San Francisco

	STR(0.1)	STR(0.2)	STR(0.3)	STR(0.4)	STR(0.5)
Failed requests	137.996	115.813	112.681	131.839	212.650
Failed rentals	70.501	61.576	62.365	74.927	104.965
Failed returns	67.495	54.237	50.316	56.912	107.685
Standard deviation	26.671	26.837	26.762	33.526	54.375
Standard error	0.843	0.849	0.846	1.060	1.719
Customer detours (min)	503.116	442.777	451.183	543.280	834.494
Relocated bikes	301.192	423.147	660.130	891.439	1361.683
Served stations	245.499	282.795	370.965	475.260	598.470
Max. runtime per k (s)	0.016	0.047	0.047	0.047	0.046
Avg. runtime per K (s)	0.045	0.045	0.047	0.051	0.061

Table B.48 Results of STR, three vehicles, and heuristic dispatching in San Francisco

	STR(0.1)	STR(0.2)	STR(0.3)	STR(0.4)	STR(0.5)
Failed requests	146.528	116.296	106.090	116.254	151.928
Failed rentals	74.190	63.584	59.011	66.509	83.365
Failed returns	72.338	52.712	47.079	49.745	68.563
Standard deviation	26.577	25.258	24.670	26.777	32.510
Standard error	0.840	0.799	0.780	0.847	1.028
Customer detours (min)	526.491	418.246	475.517	633.518	633.518
Relocated bikes	245.633	502.340	683.140	976.333	976.333
Served stations	192.970	301.063	385.838	542.444	542.444
Max. runtime per k (s)	0.047	0.033	0.047	0.047	0.047
Avg. runtime per K (s)	0.049	0.054	0.060	0.077	0.077

Table B.49 Results of SLA_{on}, three vehicles, and independent dispatching in San Francisco

	$SLA_{on}(60)$	$SLA_{on}(120)$	$SLA_{on}(180)$	$SLA_{on}(240)$	$SLA_{on}(300)$	$SLA_{on}(360)$
Failed requests	67.425	58.838	61.042	64.489	68.030	72.043
Failed rentals	46.229	38.896	40.139	42.043	44.738	47.182
Failed returns	21.196	19.942	20.903	22.446	23.292	24.861
Standard deviation	23.625	21.056	21.789	22.906	22.992	22.881
Standard error	0.747	0.666	0.689	0.724	0.727	0.724
Customer detours (min)	252.087	217.242	227.783	238.746	254.577	270.351
Relocated bikes	1122.931	1162.830	1187.040	1184.356	1168.987	1145.453
Served stations	176.695	238.494	317.145	423.899	518.870	614.090
Max. runtime per k (s)	0.141	0.187	0.540	0.543	0.357	0.973
Avg. runtime per K (s)	8.832	15.619	27.713	46.085	64.437	84.396

Table B.50 Results of SLA_{on}, three vehicles, and heuristic dispatching in San Francisco

	$SLA_{on}(60)$	$SLA_{on}(120)$	$SLA_{on}(180)$	$SLA_{on}(240)$	$SLA_{on}(300)$	$SLA_{on}(360)$
Failed requests	39.732	28.246	27.379	27.438	28.683	30.818
Failed rentals	22.591	15.596	15.402	15.669	16.632	17.914
Failed returns	17.141	12.650	11.977	11.769	12.051	12.904
Standard deviation	18.088	13.852	12.948	12.973	13.006	13.173
Standard error	0.572	0.438	0.409	0.410	0.411	0.417
Customer detours (min)	128.950	90.756	91.315	92.032	97.046	106.125
Relocated bikes	1067.145	1058.222	1045.486	1034.297	1011.560	989.363
Served stations	142.140	200.899	251.092	299.507	346.793	392.124
Max. runtime per k (s)	0.140	0.203	0.300	0.265	0.283	0.361
Avg. runtime per K (s)	8.592	13.863	21.846	31.730	42.662	54.433

Table B.51 Results of SLA_{on}, three vehicles, and optimal dispatching in San Francisco

	$SLA_{on}(60)$	$SLA_{on}(120)$	$SLA_{on}(180)$	$SLA_{on}(240)$	$SLA_{on}(300)$	$SLA_{on}(360)$
Failed requests	55.378	40.797	38.667	38.615	39.593	41.660
Failed rentals	26.484	18.120	17.230	17.603	18.544	19.926
Failed returns	28.894	22.677	21.437	21.012	21.049	21.734
Standard deviation	25.877	20.686	19.738	19.115	19.518	19.187
Standard error	0.818	0.654	0.624	0.604	0.617	0.607
Customer detours (min)	166.654	118.876	114.485	115.615	120.447	130.076
Relocated bikes	1055.109	1039.046	1021.599	994.219	975.717	944.328
Served stations	122.206	164.156	193.867	217.727	238.520	261.326
Max. runtime per k (s)	0.141	0.202	0.338	0.265	0.484	0.582
Avg. runtime per K (s)	8.954	14.628	22.563	31.907	41.832	52.779

Table B.52 Results of SLA_{off}, three vehicles and independent dispatching in San Francisco

	$SLA_{off}(60)$	$SLA_{off}(120)$	$SLA_{off}(180)$	$SLA_{off}(240)$	$SLA_{off}(300)$	$SLA_{off}(360)$
Failed requests	98.921	68.120	55.742	51.935	50.748	52.748
Failed rentals	64.634	43.764	35.546	32.797	31.541	32.591
Failed returns	34.287	24.356	20.196	19.138	19.207	20.157
Standard deviation	28.990	22.863	20.163	17.456	17.339	16.747
Standard error	0.917	0.723	0.638	0.552	0.548	0.530
Customer detours (min)	360.439	244.762	206.643	190.554	181.899	187.652
Relocated bikes	854.091	833.665	797.040	765.327	725.996	684.016
Served stations	133.999	151.518	181.451	223.724	269.100	314.000
Max. runtime per k (s)	0.047	0.063	0.063	0.063	0.062	0.063
Avg. runtime per K (s)	0.133	0.186	0.232	0.277	0.328	0.378
	$SLA_{off}(420)$	$SLA_{off}(480)$	$SLA_{off}(540)$	$SLA_{off}(600)$	$SLA_{off}(660)$	$SLA_{off}(720)$
Failed requests	57.144	61.211	66.491	70.975	73.094	73.488
Failed rentals	35.500	37.900	41.312	44.145	45.726	46.374
Failed returns	21.644	23.311	25.179	26.830	27.368	27.114
Standard deviation	18.311	18.777	20.036	20.580	20.203	20.561
Standard error	0.579	0.594	0.634	0.651	0.639	0.650
Customer detours (min)	203.119	219.854	243.485	257.420	268.317	270.565
Relocated bikes	636.827	615.051	625.149	632.347	637.240	641.193
Served stations	352.605	375.558	365.906	372.219	374.583	378.551
Max. runtime per k (s)	0.063	0.063	0.062	0.063	0.063	0.063
Avg. runtime per K (s)	0.428	0.475	0.502	0.543	0.572	0.599

Table B.53 Results of SLA_{off}, three vehicles, and heuristic dispatching in San Francisco

	$SLA_{off}(60)$	$SLA_{off}(120)$	$SLA_{off}(180)$	$SLA_{off}(240)$	$SLA_{off}(300)$	$SLA_{off}(360)$
Failed requests	73.986	49.206	40.358	36.696	34.879	35.089
Failed rentals	50.102	32.268	25.564	23.344	22.023	22.158
Failed returns	23.884	16.938	14.794	13.352	12.856	12.931
Standard deviation	23.020	18.917	16.245	15.665	14.989	15.207
Standard error	0.728	0.598	0.514	0.495	0.474	0.481
Customer detours (min)	267.729	170.827	137.294	122.822	117.701	118.735
Relocated bikes	317.669	347.246	363.196	373.251	378.730	378.987
Served stations	82.335	102.034	124.381	147.644	171.782	195.398
Max. runtime per k (s)	0.047	0.063	0.047	0.063	0.063	0.047
Avg. runtime per K (s)	0.109	0.156	0.204	0.249	0.293	0.336
	$SLA_{off}(420)$	$SLA_{off}(480)$	$SLA_{off}(540)$	$SLA_{off}(600)$	$SLA_{off}(660)$	$SLA_{off}(720)$
Failed requests	36.046	39.496	45.028	49.421	51.254	52.297
Failed rentals	23.010	25.962	31.459	36.041	37.896	38.973
Failed returns	13.036	13.534	13.569	13.380	13.358	13.324
Standard deviation	15.222	15.968	16.891	18.098	18.453	19.278
Standard error	0.481	0.505	0.534	0.572	0.584	0.610
Customer detours (min)	120.997	139.245	165.786	185.086	193.090	199.638
Relocated bikes	380.062	376.685	374.939	374.966	374.331	374.862
Served stations	219.587	240.278	245.958	243.469	244.645	245.084
Max. runtime per k (s)	0.078	0.062	0.063	0.063	0.063	0.063
Avg. runtime per K (s)	0.376	0.411	0.443	0.468	0.494	0.517

Table B.54 Results of SLA_{off}, three vehicles, and optimal dispatching in San Francisco

	$SLA_{off}(60)$	$SLA_{off}(120)$	$SLA_{off}(180)$	$SLA_{off}(240)$	$SLA_{off}(300)$	$SLA_{off}(360)$
Failed requests	75.421	53.538	44.136	40.150	39.137	39.484
Failed rentals	50.934	34.827	27.820	25.043	24.272	24.329
Failed returns	24.487	18.711	16.316	15.107	14.865	15.155
Standard deviation	25.568	22.632	20.671	18.608	17.674	17.600
Standard error	0.809	0.716	0.654	0.588	0.559	0.557
Customer detours (min)	279.512	191.264	155.308	140.454	136.144	137.196
Relocated bikes	316.605	342.910	358.062	366.830	370.165	369.666
Served stations	73.189	86.189	97.400	104.736	108.373	110.607
Max. runtime per k (s)	0.016	0.020	0.016	0.016	0.032	0.031
Avg. runtime per K (s)	0.120	0.172	0.224	0.273	0.318	0.358
	$SLA_{off}(420)$	$SLA_{off}(480)$	$SLA_{off}(540)$	$SLA_{off}(600)$	$SLA_{off}(660)$	$SLA_{off}(720)$
Failed requests	41.255	47.154	54.868	62.510	64.860	65.199
Failed rentals	25.798	30.542	37.905	44.994	47.521	47.984
Failed returns	15.457	16.612	16.963	17.516	17.339	17.215
Standard deviation	17.279	18.659	19.964	20.745	22.399	22.105
Standard error	0.546	0.590	0.631	0.656	0.708	0.699
Customer detours (min)	144.667	172.203	207.951	242.141	251.932	255.325
Relocated bikes	368.644	363.439	356.843	352.685	351.046	352.703
Served stations	110.488	109.223	100.004	95.694	95.663	95.624
Max. runtime per k (s)	0.050	0.046	0.019	0.032	0.023	0.031
Avg. runtime per K (s)	0.395	0.428	0.456	0.482	0.507	0.530

Table B.55 Results of the DLAs and three vehicles in San Francisco

	A-priori$_{off}$	A-priori$_{on}$	Ex-post$_{off}$	Ex-post$_{on}$	DLA$_{off}$	DLA$_{on}$
Failed requests	114.777	47.566	35.174	38.596	34.241	22.336
Failed rentals	64.347	26.517	21.661	22.726	21.316	12.060
Failed returns	50.430	21.049	13.513	15.870	12.925	10.276
Standard deviation	28.010	17.495	14.945	17.041	14.696	11.529
Standard error	0.886	0.553	0.473	0.539	0.465	0.365
Customer detours (min)	433.606	162.999	116.649	127.151	114.607	71.610
Relocated bikes	293.225	1059.241	383.771	1034.956	383.003	1066.104
Served stations	88.086	333.691	143.351	168.872	146.882	314.201
Max. runtime per k (s)	0.063	0.188	0.033	0.266	0.047	0.364
Avg. runtime per K (s)	0.189	22.076	0.277	11.510	0.258	27.036

Table B.56 Results of STR, four vehicles, and independent dispatching in San Francisco

	STR(0.1)	STR(0.2)	STR(0.3)	STR(0.4)	STR(0.5)
Failed requests	121.198	97.036	94.842	110.490	193.649
Failed rentals	63.465	54.006	56.747	68.863	102.650
Failed returns	57.733	43.030	38.095	41.627	90.999
Standard deviation	24.458	24.716	26.411	31.650	52.595
Standard error	0.773	0.782	0.835	1.001	1.663
Customer detours (min)	437.827	373.913	388.517	475.430	785.191
Relocated bikes	345.943	486.697	767.823	1058.185	1744.679
Served stations	294.194	348.428	476.713	623.866	805.990
Max. runtime per k (s)	0.046	0.047	0.031	0.016	0.063
Avg. runtime per K (s)	0.045	0.046	0.049	0.054	0.071

Table B.57 Results of STR, four vehicles, and heuristic dispatching in San Francisco

	STR(0.1)	STR(0.2)	STR(0.3)	STR(0.4)	STR(0.5)
Failed requests	136.576	100.363	83.632	87.256	116.488
Failed rentals	70.839	57.735	50.126	56.566	74.910
Failed returns	65.737	42.628	33.506	30.690	41.578
Standard deviation	25.393	22.795	22.368	23.120	27.627
Standard error	0.803	0.721	0.707	0.731	0.874
Customer detours (min)	488.078	332.057	370.918	518.494	518.494
Relocated bikes	255.089	524.660	720.143	1089.941	1089.941
Served stations	199.570	340.280	455.087	696.346	696.346
Max. runtime per k (s)	0.063	0.047	0.047	0.047	0.047
Avg. runtime per K (s)	0.051	0.059	0.069	0.101	0.101

Table B.58 Results of SLA_{on} four vehicles, and independent dispatching in San Francisco

	$SLA_{on}(60)$	$SLA_{on}(120)$	$SLA_{on}(180)$	$SLA_{on}(240)$	$SLA_{on}(300)$	$SLA_{on}(360)$
Failed requests	54.354	51.971	53.826	57.270	61.198	65.799
Failed rentals	40.109	36.385	36.903	39.100	42.220	45.205
Failed returns	14.245	15.586	16.923	18.170	18.978	20.594
Standard deviation	20.532	19.125	19.733	19.451	20.872	20.090
Standard error	0.649	0.605	0.624	0.615	0.660	0.635
Customer detours (min)	209.684	194.102	201.053	212.837	227.763	244.785
Relocated bikes	1425.438	1462.282	1490.671	1491.500	1471.155	1440.240
Served stations	243.536	325.603	413.649	530.586	640.509	747.497
Max. runtime per k (s)	0.156	0.218	1.747	0.460	0.346	0.524
Avg. runtime per K (s)	13.227	23.276	39.052	62.154	86.316	110.869

Table B.59 Results of SLA_{on}, four vehicles, and heuristic dispatching in San Francisco

	$SLA_{on}(60)$	$SLA_{on}(120)$	$SLA_{on}(180)$	$SLA_{on}(240)$	$SLA_{on}(300)$	$SLA_{on}(360)$
Failed requests	20.369	14.983	15.057	14.800	15.312	16.187
Failed rentals	11.907	8.783	9.053	8.934	9.455	10.214
Failed returns	8.462	6.200	6.004	5.866	5.857	5.973
Standard deviation	11.699	8.954	9.229	8.315	8.239	8.154
Standard error	0.370	0.283	0.292	0.263	0.261	0.258
Customer detours (min)	61.654	47.035	49.645	48.538	50.463	55.137
Relocated bikes	1346.802	1295.274	1259.589	1239.625	1221.163	1192.421
Served stations	179.502	270.792	339.298	391.533	440.342	491.583
Max. runtime per k (s)	0.156	0.203	0.796	0.362	1.545	1.282
Avg. runtime per K (s)	13.123	21.411	32.742	45.921	60.286	76.084

Table B.60 Results of SLA_{on}, four vehicles, and optimal dispatching in San Francisco

	$SLA_{on}(60)$	$SLA_{on}(120)$	$SLA_{on}(180)$	$SLA_{on}(240)$	$SLA_{on}(300)$	$SLA_{on}(360)$
Failed requests	37.788	27.630	26.286	27.324	28.702	28.947
Failed rentals	16.559	11.202	10.527	11.381	12.319	12.801
Failed returns	21.229	16.428	15.759	15.943	16.383	16.146
Standard deviation	24.251	18.302	17.395	18.094	18.679	18.536
Standard error	0.767	0.579	0.550	0.572	0.591	0.586
Customer detours (min)	104.184	73.356	71.435	75.591	80.321	83.513
Relocated bikes	1321.472	1278.456	1230.093	1202.886	1181.390	1149.902
Served stations	148.090	215.429	259.599	283.770	302.505	322.961
Max. runtime per k (s)	0.172	0.219	0.265	0.312	0.435	0.794
Avg. runtime per K (s)	13.647	22.761	34.625	47.760	61.731	76.612

Table B.61 Results of SLA_{off}, four vehicles, and independent dispatching in San Francisco

	$SLA_{off}(60)$	$SLA_{off}(120)$	$SLA_{off}(180)$	$SLA_{off}(240)$	$SLA_{off}(300)$	$SLA_{off}(360)$
Failed requests	87.269	58.059	47.754	45.548	46.807	46.782
Failed rentals	61.052	39.371	31.586	29.438	29.979	29.795
Failed returns	26.217	18.688	16.168	16.110	16.828	16.987
Standard deviation	27.924	19.961	17.454	15.402	15.425	15.474
Standard error	0.883	0.631	0.552	0.487	0.488	0.489
Customer detours (min)	325.311	211.944	176.750	165.336	163.725	161.927
Relocated bikes	700.518	705.737	714.588	731.209	731.747	736.350
Served stations	174.224	197.007	237.213	291.523	350.935	412.675
Max. runtime per k (s)	0.047	0.062	0.063	0.062	0.063	0.063
Avg. runtime per K (s)	0.157	0.235	0.311	0.395	0.481	0.573
	$SLA_{off}(420)$	$SLA_{off}(480)$	$SLA_{off}(540)$	$SLA_{off}(600)$	$SLA_{off}(660)$	$SLA_{off}(720)$
Failed requests	50.403	55.692	62.671	66.772	68.669	68.509
Failed rentals	32.563	36.379	40.794	43.585	45.068	45.474
Failed returns	17.840	19.313	21.877	23.187	23.601	23.035
Standard deviation	16.369	17.369	18.707	19.396	19.334	19.150
Standard error	0.518	0.549	0.592	0.613	0.611	0.606
Customer detours (min)	176.288	198.819	227.234	241.185	249.651	251.378
Relocated bikes	727.026	707.183	711.121	719.901	720.770	720.731
Served stations	463.586	483.838	480.834	487.171	493.490	500.062
Max. runtime per k (s)	0.065	0.063	0.063	0.078	0.062	0.063
Avg. runtime per K (s)	0.667	0.732	0.791	0.852	0.902	0.947

Table B.62 Results of SLA_{off}, four vehicles, and heuristic dispatching in San Francisco

	$SLA_{off}(60)$	$SLA_{off}(120)$	$SLA_{off}(180)$	$SLA_{off}(240)$	$SLA_{off}(300)$	$SLA_{off}(360)$
Failed requests	63.566	42.022	35.885	32.691	31.809	31.841
Failed rentals	45.469	28.981	24.340	22.168	21.196	21.504
Failed returns	18.097	13.041	11.545	10.523	10.613	10.337
Standard deviation	21.451	17.449	15.830	15.308	14.605	14.671
Standard error	0.678	0.552	0.501	0.484	0.462	0.464
Customer detours (min)	233.733	147.561	124.538	112.952	110.214	110.506
Relocated bikes	335.023	362.060	376.771	389.123	393.238	395.126
Served stations	86.328	109.038	132.100	159.224	187.893	215.555
Max. runtime per k (s)	0.062	0.047	0.062	0.063	0.063	0.063
Avg. runtime per K (s)	0.150	0.231	0.310	0.388	0.462	0.534
	$SLA_{off}(420)$	$SLA_{off}(480)$	$SLA_{off}(540)$	$SLA_{off}(600)$	$SLA_{off}(660)$	$SLA_{off}(720)$
Failed requests	32.053	35.569	41.388	47.284	49.956	51.620
Failed rentals	21.620	25.143	30.737	36.822	39.418	40.967
Failed returns	10.433	10.426	10.651	10.462	10.538	10.653
Standard deviation	14.670	15.669	17.066	17.073	18.098	18.788
Standard error	0.464	0.495	0.540	0.540	0.572	0.594
Customer detours (min)	111.492	128.129	157.225	184.107	192.373	201.600
Relocated bikes	396.626	395.554	393.674	392.869	391.463	391.515
Served stations	248.144	278.240	291.184	294.002	293.218	293.142
Max. runtime per k (s)	0.063	0.063	0.063	0.063	0.078	0.078
Avg. runtime per K (s)	0.601	0.666	0.717	0.764	0.805	0.843

Table B.63 Results of SLA_{off}, four vehicles, and optimal dispatching in San Francisco

	$SLA_{off}(60)$	$SLA_{off}(120)$	$SLA_{off}(180)$	$SLA_{off}(240)$	$SLA_{off}(300)$	$SLA_{off}(360)$
Failed requests	67.427	47.139	40.939	38.459	36.333	36.924
Failed rentals	47.433	31.320	25.686	23.534	22.320	22.750
Failed returns	19.994	15.819	15.253	14.925	14.013	14.174
Standard deviation	23.349	21.577	20.062	19.765	18.252	18.320
Standard error	0.738	0.682	0.634	0.625	0.577	0.579
Customer detours (min)	250.689	168.218	142.490	132.427	126.123	128.389
Relocated bikes	334.513	359.197	373.206	382.775	386.593	386.771
Served stations	74.651	89.724	101.444	110.282	115.107	118.652
Max. runtime per k (s)	0.020	0.016	0.031	0.020	0.031	0.022
Avg. runtime per K (s)	0.165	0.254	0.339	0.420	0.494	0.564
	$SLA_{off}(420)$	$SLA_{off}(480)$	$SLA_{off}(540)$	$SLA_{off}(600)$	$SLA_{off}(660)$	$SLA_{off}(720)$
Failed requests	37.203	41.304	48.435	54.598	57.510	58.366
Failed rentals	23.677	27.265	34.283	40.501	43.211	44.216
Failed returns	13.526	14.039	14.152	14.097	14.299	14.150
Standard deviation	17.683	18.541	19.539	21.516	21.872	21.608
Standard error	0.559	0.586	0.618	0.680	0.692	0.683
Customer detours (min)	131.010	149.644	183.122	208.477	220.537	225.803
Relocated bikes	388.125	384.252	380.484	377.728	375.092	375.071
Served stations	120.302	120.547	112.954	108.785	107.842	107.717
Max. runtime per k (s)	0.031	0.031	0.031	0.036	0.032	0.027
Avg. runtime per K (s)	0.626	0.683	0.730	0.775	0.816	0.855

Table B.64 Results of the DLAs and four vehicles in San Francisco

	A-priori$_{off}$	A-priori$_{on}$	Ex-post$_{off}$	Ex-post$_{on}$	DLA$_{off}$	DLA$_{on}$
Failed requests	107.933	33.442	30.581	21.625	30.204	13.217
Failed rentals	63.185	21.031	20.351	13.033	20.276	7.210
Failed returns	44.748	12.411	10.230	8.592	9.928	6.007
Standard deviation	28.126	13.908	14.016	11.774	14.019	8.687
Standard error	0.889	0.440	0.443	0.372	0.443	0.275
Customer detours (min)	416.008	117.548	104.376	67.667	104.417	39.699
Relocated bikes	309.914	1332.609	401.560	1288.869	399.483	1292.965
Served stations	88.504	414.787	169.478	213.817	160.351	333.344
Max. runtime per k (s)	0.063	1.123	0.062	0.406	0.062	0.369
Avg. runtime per K (s)	0.283	32.169	0.467	17.998	0.394	34.842

References

Aarts E, Korst J (1989) Simulated annealing and Boltzmann machines. Wiley, New York

Alvarez-Valdes R, Belenguer JM, Benavent E, Bermudez JD, Muñoz F, Vercher E, Verdejo F (2016) Optimizing the level of service quality of a bike-sharing system. Omega 64:163–175. https://doi.org/10.1016/j.omega.2015.09.007

Ansmann A, Ulmer MW, Mattfeld DC (2017) Spatial information in offline approximate dynamic programming for dynamic vehicle routing with stochastic requests. Logist Manag 2017:25–32

Arabzad SM, Shirouyehzad H, Bashiri M, Tavakkoli-Moghaddam R, Najafi E (2018) Rebalancing static bike-sharing systems: a two-period two-commodity multi-depot mathematical model. Transport 33:718–726. https://doi.org/10.3846/transport.2018.1558

Archer G (2017) Does sharing cars really reduce car use? Technical report, Transport & Environment,. https://www.transportenvironment.org/sites/te/files/publications/Does-sharing-cars-really-reduce-car-use-June%202017.pdf. Accessed 7 Mar 2018

Bazaar MS, Jarvis JJ, Sherali HD (2010) Linear programming and network flows. John Wiley & Sons Inc, Hoboken

Bellman R (1957) A Markovian decision process. Technical report, DTIC Document

Borgnat P, Abry P, Flandrin P, Robardet C, Rouquier JB, Fleury E (2011) Shared bicycles in a city: a signal processing and data analysis perspective. Adv Complex Syst 14(3):415–438. https://doi.org/10.1142/S0219525911002950

Brinkmann J, Ulmer MW, Mattfeld DC (2015) Short-term strategies for stochastic inventory routing in bike sharing systems. Transp Res Proc 10:364–373. https://doi.org/10.1016/j.trpro.2015.09.086

Brinkmann J, Ulmer MW, Mattfeld DC (2016) Inventory routing for bikes sharing systems. Transp Res Proc 19:316–327. https://doi.org/10.1016/j.trpro.2016.12.091

Brinkmann J, Ulmer MW, Mattfeld DC (2019a) Dynamic lookahead policies for stochastic-dynamic inventory routing in bike sharing systems. Comput Oper Res 106:260–279. https://doi.org/10.1016/j.cor.2018.06.004

Brinkmann J, Ulmer MW, Mattfeld DC (2019b) The multi-vehicle stochastic-dynamic inventory routing problem for bike sharing systems. Bus Res. https://doi.org/10.1007/s40685-019-0100-z

Bulhões T, Subramanian A, Erdoğan G, Laporte G (2018) The static bike relocation problem with multiple vehicles and visits. Eur J Oper Res 264(2):508–523. https://doi.org/10.1016/j.ejor.2017.06.028

Büttner J, Mlasowsky H, Birkholz T et al (2011) Optimising bike sharing in European cities–a handbook. OBIS project

J. Brinkmann, *Active Balancing of Bike Sharing Systems*, Lecture Notes in Mobility,
https://doi.org/10.1007/978-3-030-35012-3

Butz MV, Sigaud O, Gérard P (2003) Anticipatory behavior: exploiting knowledge about the future to improve current behavior. Anticipatory behavior in adaptive learning systems: foundations, theories, and systems. Springer, Berlin, pp 1–10. https://doi.org/10.1007/978-3-540-45002-3_1

Caggiani L, Ottomanelli M (2012) A modular soft computing based method for vehicles repositioning in bike-sharing systems. Proc-Soc Behav Sci 54:675–684. https://doi.org/10.1016/j.sbspro.2012.09.785 Elsevier

Caggiani L, Ottomanelli M (2013) A dynamic simulation based model for optimal fleet repositioning in bike-sharing systems. Proc-Soc Behav Sci 87:203–210. https://doi.org/10.1016/j.sbspro.2013.10.604

Çelebi D, Yörüsün A, Işık H (2018) Bicycle sharing system design with capacity allocations. Transp Res Part B: Methodol 118:86–98. https://doi.org/10.1016/j.trb.2018.05.018

Chemla D, Meunier F, Wolfler Calvo R (2013) Bike sharing systems: solving the static rebalancing problem. Discret Optim 10(2):120–146. https://doi.org/10.1016/j.disopt.2012.11.005

Chiariotti F, Pielli C, Zanella A, Zorzi M (2018) A dynamic approach to rebalancing bike-sharing systems. Sensors 18(2):512. https://doi.org/10.3390/s18020512

Coelho LC, Cordeau JF, Laporte G (2014a) Heuristics for dynamic and stochastic inventory-routing. Comput Oper Res 52(A):55–67. https://doi.org/10.1016/j.cor.2014.07.001

Coelho LC, Cordeau JF, Laporte G (2014b) Thirty years of inventory routing. Transp Sci 48(1):1–19. https://doi.org/10.1287/trsc.2013.0472

Contardo C, Morency C, Rousseau LM (2012) Balancing a dynamic public bike-sharing system. In: CIRRELT-2012-09. https://www.cirrelt.ca/DocumentsTravail/CIRRELT-2012-09.pdf. Accessed 8 Dec 2014

Crainic TG, Laporte G (1997) Planning models for freight transportation. Eur J Oper Res 97:409–438. https://doi.org/10.1016/S0377-2217(96)00298-6

Cruz F, Subramanian A, Bruck BP, Iori M (2017) A heuristic algorithm for a single vehicle static bike sharing rebalancing problem. Comput Oper Res 79:19–33. https://doi.org/10.1016/j.cor.2016.09.025

Datner S, Raviv T, Tzur M, Chemla D (2017) Setting inventory levels in a bike sharing network. Transp Sci 53(1):62–76. https://doi.org/10.1287/trsc.2017.0790

Dell'Amico M, Hadjicostantinou E, Iori M, Novellani S (2014) The bike sharing rebalancing problem: mathematical formulations and benchmark instances. Omega 45:7–19. https://doi.org/10.1016/j.omega.2013.12.001

Dell'Amico M, Iori M, Novellani S, Stützle T (2016) A destroy and repair algorithm for the Bike sharing rebalancing problem. Comput Oper Res 71:149–162. https://doi.org/10.1016/j.cor.2016.01.011

Dell'Amico M, Iori M, Novellani S, Subramanian A (2018) The bike sharing rebalancing problem with stochastic demands. Transp Res Part B: Methodol 118:362–380. https://doi.org/10.1016/j.trb.2018.10.015

DeMaio P (2009) Bike-sharing: history, impacts, models of provision, and future. J Public Transp 12(4):3. https://doi.org/10.5038/2375-0901.12.4.3

Deriyenko T, Hartkopp O, Mattfeld DC (2017) Supporting product optimization by customer data analysis. In: Operations Research Proceedings 2015,. Springer, Cham, pp 491–496. https://doi.org/10.1007/978-3-319-42902-1_66

Di Gaspero L, Rendl A, Urli T (2013a) A hybrid ACO+CP for balancing bicycle sharing systems. Hybrid metaheuristics, vol 7919. Lecture notes in computer science. Springer, Berlin, pp 198–212. https://doi.org/10.1007/978-3-642-38516-2_16

Di Gaspero L, Rendl A, Urli T (2013b) Constraint-based approaches for balancing bike sharing systems. Principles and practice of constraint programming, vol 8124. Lecture notes in computer science. Springer, Berlin, pp 758–773. https://doi.org/10.1007/978-3-642-40627-0_56

Dong L, Liang H, Zhang L, Liu Z, Gao Z, Hu M (2017) Highlighting regional eco-industrial development: Life cycle benefits of an urban industrial symbiosis and implications in China. Ecol Model 361:164–176. https://doi.org/10.1016/j.ecolmodel.2017.07.032

Erdoğan G, Laporte G, Calvo RW (2014) The static bicycle relocation problem with demand intervals. Eur J Oper Res 238(2):451–457. https://doi.org/10.1016/j.ejor.2014.04.013

Erdoğan G, Battarra M, Calvo RW (2015) An exact algorithm for the static rebalancing problem arising in bicycle sharing systems. Eur J Oper Res 245(3):667–679. https://doi.org/10.1016/j.ejor.2015.03.043

Espegren HM, Kristianslund J, Andersson H, Fagerholt K (2016) The static bicycle repositioning problem - literature survey and new formulation. Computational logistics, vol 9855. Lecture notes in computer science. Springer, Cham, pp 337–351. https://doi.org/10.1007/978-3-319-44896-1_22

Fishman E, Washington S, Haworth N (2014) Bike share's impact on car use: evidence from the United States, Great Britain, and Australia. Transp Res Part D: Transp Environ 31:13–20. https://doi.org/10.1016/j.trd.2014.05.013

Forma IA, Raviv T, Tzur M (2015) A 3-step math heuristic for the static repositioning problem in bike-sharing systems. Transp Res Part B: Methodol 71:230–247. https://doi.org/10.1016/j.trb.2014.10.003

Fricker C, Gast N (2016) Incentives and redistribution in homogeneous bike-sharing systems with stations of finite capacity. EURO J Transp Logist 5(3):261–291. https://doi.org/10.1007/s13676-014-0053-5

Froehlich J, Neumann J, Oliver N (2009) Sensing and predicting the pulse of the city through shared bicycling. In: Proceedings of the twenty-first international joint conference on artificial intelligence (IJCAI-09), pp 1420–1426. https://www.ijcai.org/Proceedings/2009/. Accessed 13 Feb 2017

García-Palomares JC, Gutiérrez J, Latorre M (2012) Optimizing the location of stations in bike-sharing programs: a GIS approach. Appl Geogr 35(1):235–246. https://doi.org/10.1016/j.apgeog.2012.07.002

Gauthier A, Hughes C, Kost C, Li S, Linke C, Lotshaw S, Mason J, Pardo C, Rasore C, Schroeder B, Treviño X (2013) The bike-share planning guide. Technical report, Institute for Transportation and Development Policy. https://www.itdp.org/wp-content/uploads/2014/07/ITDP_Bike_Share_Planning_Guide.pdf. Accessed 31 July 2014

Ghiani G, Manni E, Quaranta A, Triki C (2009) Anticipatory algorithms for same-day courier dispatching. Transp Res Part E: Logist Transp Rev 45(1):96–106. https://doi.org/10.1016/j.tre.2008.08.003

Ghosh S, Trick MA, Varakantham P (2016) Robust repositioning to counter unpredictable demand in bike sharing systems. In: Proceedings of the twenty-fifth international joint conference on artificial intelligence (IJCAI-16). http://www.ijcai.org/Proceedings/2016. Accessed 11 Dec 2017

Ghosh S, Varakantham P, Adulyasak Y, Jaillet P (2017) Dynamic Repositioning to Reduce Lost Demand in Bike Sharing Systems. J Artif Intell Res 58:387–430. https://doi.org/10.1613/jair.5308

Goodson JC, Thomas BW, Ohlmann JW (2017) A Rollout Algorithm Framework for Heuristic Solutions to Finite-Horizon Stochastic Dynamic Programs. Eur J Oper Res 258(1):216–229. https://doi.org/10.1016/j.ejor.2016.09.040

Groß PO, Ehmke JF, Haas I, Mattfeld DC (2017) Evaluation of alternative paths for reliable routing in city logistics. Transp Res Proc 27:1,195–1,202. https://doi.org/10.1016/j.trpro.2017.12.067

Hermanns JAL, Brinkmann J, Mattfeld DC (2019) Dynamic policy selection for a stochastic-dynamic knapsack problem. In: Operations research proceedings 2018. https://doi.org/10.1007/978-3-030-18500-8_40

Ho SC, Szeto WY (2014) Solving a static repositioning problem in bike-sharing systems using iterated tabu search. Transp Res Part E: Logist Transp Rev 69:180–198. https://doi.org/10.1016/j.tre.2014.05.017

Ho SC, Szeto WY (2017) A hybrid large neighborhood search for the static multi-vehicle bike-repositioning problem. Transp Res Part B: Methodol 95:340–363. https://doi.org/10.1016/j.trb.2016.11.003

Hofmann-Wellenhof B, Lichtenegger H, Collins J (2012) Global positioning system: theory and practice, 4th edn. Springer, Cham

Houghton JT, Jenkins GJ, Ephraums JJ (1990) Climate change: the IPCC scientific assessment. Cambridge University Press, Cambridge

Kadri AA, Labadi K, Kacem I (2015) An integrated petri net and genetic algorithm based approach for performance optimization of bicycle-sharing systems. Eur J Ind Eng 9(5):638–663. https://doi.org/10.1504/EJIE.2015.071777

Kadri AA, Kacem I, Labadi K (2016) A branch-and-bound algorithm for solving the static rebalancing problem in bicycle-sharing systems. Comput Ind Eng 95:42–52. https://doi.org/10.1016/j.cie.2016.02.002

Kadri AA, Kacem I, Labadi K (2018) Lower and upper bounds for scheduling multiple balancing vehicles in bicycle-sharing systems. Soft Comput. https://doi.org/10.1007/s00500-018-3258-y

Kall P, Wallace SW (1994) Stochastic programming. Wiley, New York

Kaltenbrunner A, Meza R, Grivolla J, Codina J, Banchs R (2010) Urban cycles and mobility patterns: exploring and predicting trends in a bicycle-based public transport system. Pervasive Mob Comput 6(4):455–466. https://doi.org/10.1016/j.pmcj.2010.07.002

Kaspi M, Raviv T, Tzur M (2014) Parking reservation policies in one-way vehicle sharing systems. Transp Res Part B: Methodol 62:35–50. https://doi.org/10.1016/j.trb.2014.01.006

Kaspi M, Raviv T, Tzur M (2016) Detection of unusable bicycles in bike-sharing systems. Omega 65:10–16. https://doi.org/10.1016/j.omega.2015.12.003

Kloimüllner C, Raidl GR (2017a) Full-load route planning for balancing bike sharing systems by logic-based benders decomposition. Networks 69(3):270–289. https://doi.org/10.1002/net.21736

Kloimüllner C, Raidl GR (2017b) Hierarchical clustering and multilevel refinement for the bike-sharing station planning problem. Learning and intelligent optimization, vol 10,556. Lecture notes in computer science. Springer, Cham, pp 150–165. https://doi.org/10.1007/978-3-319-69404-7_11

Kloimüllner C, Papazek P, Hu B, Raidl GR (2014) Balancing bicycle sharing systems: an approach for the dynamic case. Evolutionary computation in combinatorial optimization, vol 8600. Lecture notes in computer science. Springer, Berlin, pp 73–84. https://doi.org/10.1007/978-3-662-44320-0_7

Kloimüllner C, Papazek P, Hu B, Raidl GR (2015) A cluster-first route-second approach for balancing bicycle sharing systems. In: Computer aided systems theory - EUROCAST 2015. Lecture notes in computer science, vol 9520. Springer, Cham, pp 439–446. https://doi.org/10.1007/978-3-319-27340-2_55

Köster F, Ulmer MW, Mattfeld DC, Hasle G (2018) Anticipating emission-sensitive traffic management strategies for dynamic delivery routing. Transp Res Part D: Transp Environ 62:345–361. https://doi.org/10.1016/j.trd.2018.03.002

Laporte G, Mercure H, Nobert Y (1986) An exact algorithm for the asymmetrical capacitated vehicle routing problem. Networks 16(1):33–46. https://doi.org/10.1002/net.3230160104

Legros B (2019) Dynamic repositioning strategy in a bike-sharing system; how to prioritize and how to rebalance a bike station. Eur J Oper Res 272(2):740–753. https://doi.org/10.1016/j.ejor.2018.06.051

Li Y, Szeto WY, Long J, Shui CS (2016) A multiple type bike repositioning problem. Transp Res Part B: Methodol 90:263–278. https://doi.org/10.1016/j.trb.2016.05.010

Lin JR, Yang TH (2011) Strategic design of public bicycle sharing systems with service level constraints. Transp Res Part E: Logist Transp Rev 47(2):284–294. https://doi.org/10.1016/j.tre.2010.09.004

Lu CC (2016) Robust multi-period fleet allocation models for bike-sharing systems. Netw Spat Econ 16(1):61–82. https://doi.org/10.1007/s11067-013-9203-9

Manna C (2016) On-line dynamic station redeployments in bike-sharing systems. In: AI*IA 2016 advances in artificial intelligence. Lecture notes in computer science, vol 10,037. Springer, Cham, pp 13–25. https://doi.org/10.1007/978-3-319-49130-1_2

Markowitz H (1952) Portfolio selection. J Financ 7(1):77–91. https://doi.org/10.1111/j.1540-6261.1952.tb01525.x

Massink R, Zuidgeest M, Rijnsburger J, Sarmiento OL, van Maarseveen M (2011) The climate value of cycling. Nat Resour Forum 35:100–111. https://doi.org/10.1111/j.1477-8947.2011.01345.x

Mattfeld D, Vahrenkamp R (2014) Logistiknetzwerke: Modelle für Standortwahl und Tourenplanung. Springer

Meddin R (2017) The bike-sharing map. Twitter, Inc. https://twitter.com/BikesharingMap/status/928600813047566336. Accessed 9 Nov 2017

Meisel S (2011) Anticipatory optimization for dynamic decision making. Operations research, vol 51. Computer science interfaces series. Springer, New York. https://doi.org/10.1007/978-1-4614-0505-4

Midgley P (2009) The role of smart bike-sharing systems in urban mobility. Journeys 2:23–31

Miller CE, Tucker AW, Zemlin RA (1960) Integer programming formulation of traveling salesman problems. J Assoc Comput Mach 7(4):326–329. https://doi.org/10.1145/321043.321046

Motivate International Inc (2015) Bay area bike share, San Francisco, CA, USA. https://www.fordgobike.com/. Accessed 21 Dec 2015

Motivate International Inc (2016) Nice ride, Minneapolis, MN, USA. https://www.niceridemn.org/. Accessed 31 Aug 2016

Motivate International Inc (2018) City bike, New York City, NY, USA. https://www.citibikenyc.com/. Accessed 5 July 2018

Neumann Saavedra BA (2018) Service network design of bike sharing systems with resource-management consideration. PhD thesis, Technische Universität Braunschweig. https://katalog.ub.tu-braunschweig.de/vufind/Record/1015689094

Neumann Saavedra BA, Vogel P, Mattfeld DC (2015) Anticipatory service network design of bike sharing systems. Transp Res Proc 10:355–363. https://doi.org/10.1016/j.trpro.2015.09.085

Neumann Saavedra BA, Crainic TG, Gendron B, Mattfeld DC, Römer M (2016) Service network design of bike sharing systems with resource constraints. Computational logistics, vol 9855. Lecture notes in computer science. Springer, Cham, pp 352–366. https://doi.org/10.1007/978-3-319-44896-1_23

O'Brien O, Cheshire J, Batty M (2014) Mining bicycle sharing data for generating insights into sustainable transport systems. J Transp Geogr 34:262–273. https://doi.org/10.1016/j.jtrangeo.2013.06.007

Pal A, Zhang Y (2017) Free-floating bike sharing: solving real-life large-scale static rebalancing problems. Transp Res Part C: Emerg Technol 80:92–116. https://doi.org/10.1016/j.trc.2017.03.016

Papazek P, Raidl GR, Rainer-Harbach M, Hu B (2013) A PILOT/VND/GRASP hybrid for the static balancing of public bicycle sharing systems. In: Computer aided systems theory - EUROCAST 2013: 14th international conference. Lecture notes in computer science. Springer, Berlin, pp 372–379. https://doi.org/10.1007/978-3-642-53856-8_47

Papazek P, Kloimüllner C, Hu B, Raidl GR (2014) Balancing bicycle sharing systems: an analysis of path relinking and recombination within a GRASP hybrid. Parallel Problem solving from nature – PPSN XIII, vol 8672. Lecture notes in computer science. Springer, Cham, pp 792–801. https://doi.org/10.1007/978-3-319-10762-2_78

Park C, Sohn SY (2017) An optimization approach for the placement of bicycle-sharing stations to reduce short car trips: an application to the city of Seoul. Transp Res Part A: Policy Pract 105:154–166. https://doi.org/10.1016/j.tra.2017.08.019

Powell WB (2011) Approximate dynamic programming: solving the curses of dimensionality, 2nd edn. Wiley, New York

Powell WB, Towns MT, Marar A (2000) On the value of optimal myopic solutions for dynamic routing and scheduling problems in the presence of user noncompliance. Transp Sci 34(1):67–85. https://doi.org/10.1287/trsc.34.1.67.12283

Pratte J (2006) Bicycle tourism: on the trail to economic development. Prairie perspectives: geographical essays 9(1):65–84 Department of Geography, University of Manitoba, Winnipeg, Manitoba, Canada

Puterman ML (2014) Markov decision processes: discrete stochastic dynamic programming, 2nd edn. Wiley, New York

Rabl A, de Nazelle A (2012) Benefits of shift from car to active transport. Transp Policy 19(1):121–131. https://doi.org/10.1016/j.tranpol.2011.09.008

Rainer-Harbach M, Papazek P, Hu B, Raidl GR (2013) Balancing bicycle sharing systems: a variable neighborhood search approach. Evolutionary Computation in combinatorial optimization, vol 7832. Lecture notes in computer science. Springer, Berlin, pp 121–132. https://doi.org/10.1007/978-3-642-37198-1_11

Rainer-Harbach M, Papazek P, Raidl GR, Hu B, Kloimüllner C (2015) PILOT, GRASP, and VNS approaches for the static balancing of bicycle sharing systems. J Glob Optim 63(3):597–629. https://doi.org/10.1007/s10898-014-0147-5

Raviv T, Kolka O (2013) Optimal inventory management of a bike-sharing station. IIE Trans 45(10):1077–1093. https://doi.org/10.1080/0740817X.2013.770186

Raviv T, Tzur M, Forma IA (2013) Static repositioning in a bike-sharing system: models and solution approaches. EURO J Transp Logist 2(3):187–229. https://doi.org/10.1007/s13676-012-0017-6

Reiss S, Bogenberger K (2015) GPS-data analysis of Munich's free-floating bike sharing system and application of an operator-based relocation strategy. In: 2015 IEEE 18th international conference on intelligent transportation systems, pp 584–589. https://doi.org/10.1109/ITSC.2015.102

Reiss S, Bogenberger K (2016) Optimal bike fleet management by smart relocation methods: combining an operator-based with an user-based relocation strategy. In: 2016 IEEE 19th international conference on intelligent transportation systems (ITSC), pp 2613–2618. https://doi.org/10.1109/ITSC.2016.7795976

Ricker VHH (2015) Ein Dekompositionsansatz für das taktisch-operative management von bike-sharing-systemen. PhD thesis, Technische Universität Braunschweig. http://www.digibib.tu-bs.de/?docid=00062118

Rodigue JP, Comtois C, Slack B (2013) The geography of transport systems, 3rd edn. Routledge, London

Rosenkrantz D, Stearns R, Lewis P II (1977) An analysis of several heuristics for the traveling salesman problem. SIAM J Comput 6(3):563–581. https://doi.org/10.1137/0206041

Rothlauf F (2011) Design of modern heuristics: priciples and applications. Springer, Berlin. https://doi.org/10.1007/978-3-540-72962-4

Rudloff C, Lackner B (2014) Modeling demand for bikesharing systems - neighboring stations as source for demand and reason for structural breaks. Transp Res Rec: J Transp Res Board 2430:1–11. https://doi.org/10.3141/2430-01

Savelsbergh MWP, Sol M (1995) The general pickup and delivery problem. Transp Sci 29(1):17–29. https://doi.org/10.1287/trsc.29.1.17

Scherr YO, Neumann Saavedra BA, Hewitt M, Mattfeld DC (2018) Service network design for same day delivery with mixed autonomous fleets. Transp Res Proc 30:23–32. https://doi.org/10.1016/j.trpro.2018.09.004

Schuijbroek J, Hampshire RC, van Hoeve WJ (2017) Inventory rebalancing and vehicle routing in bike sharing systems. Eur J Oper Res 257(3):992–1004. https://doi.org/10.1016/j.ejor.2016.08.029

Shaheen S, Guzman S, Zhang H (2010) Bikesharing in Europe, the Americas, and Asia: past, present, and future. Transp Res Rec: J Transp Res Board 2143:159–167. https://doi.org/10.3141/2143-20

Shaheen S, Zhang H, Martin E, Guzman S (2011) Hangzhou public bicycle: understanding early adoption and behavioral response to bikesharing in Hangzhou, China. Transp Res Rec 2247:34–41. https://escholarship.org/uc/item/31510910. Accessed 1 June 2018

Shui CS, Szeto WY (2017) Dynamic green bike repositioning problem - a hybrid rolling horizon artificial bee colony algorithm approach. Transp Res Part D: Transp Environ. https://doi.org/10.1016/j.trd.2017.06.023

Sieg G (2008) Volkswirtschaftslehre - Mit aktuellen Fallstudien, 2nd edn. Oldenbourg

Singla A, Santoni M, Bartók G, Mukerji P, Meenen M, Krause A (2015) Incentivizing users for balancing bike sharing systems. In: Proceedings of the twenty-ninth AAAI conference on artificial intelligence

Soeffker N, Ulmer MW, Mattfeld DC (2016) Problem-specific state space partitioning for dynamic vehicle routing problems. In: Proceedings of MKWI, pp 229–240

Soeffker N, Ulmer MW, Mattfeld DC (2017) On fairness aspects of customer acceptance mechanisms in dynamic vehicle routing. Logist Manag 2017:17–24

Solomon MM (1987) Algorithms for the vehicle routing and scheduling problems with time window constraints. Oper Res 35(2):254–265. https://doi.org/10.1287/opre.35.2.254

Stern KL (2012) Hungarian algorithm. GitHub, Inc. https://github.com/KevinStern/software-and-algorithms/blob/master/src/main/java/blogspot/software_and_algorithms/stern_library/optimization/HungarianAlgorithm.java. Accessed 15 Oct 2018

Szeto WY, Shui CS (2018) Exact loading and unloading strategies for the static multi-vehicle bike repositioning problem. Transp Res Part B: Methol 109:176–211. https://doi.org/10.1016/j.trb.2018.01.007

Szeto WY, Liu Y, Ho SC (2016) Chemical reaction optimization for solving a static bike repositioning problem. Transp Res Part D: Transp Environ 47:104–135. https://doi.org/10.1016/j.trd.2016.05.005

Tang Q, Fu Z, Qiu M (2019) A bilevel programming model and algorithm for the static bike repositioning problem. J Adv Transp. https://doi.org/10.1155/2019/8641492

Toth P, Vigo D (2014) Vehicle routing: problems, methods, and applications, 2nd edn. Society for Industrial and Applied Mathematics, Philadelphia

Ulmer MW (2017) Approximate dynamic programming for dynamic vehicle routing. Operations research, vol 61. Computer science interfaces series. Springer, Berlin. https://doi.org/10.1007/978-3-319-55511-9

Ulmer MW, Voß S (2016) Risk-averse anticipation for dynamic vehicle routing. In: International conference on learning and intelligent optimization. Lecture notes in computer science, vol 10079. Springer, Cham, pp 274–279. https://doi.org/10.1007/978-3-319-50349-3_23

Ulmer MW, Brinkmann J, Mattfeld DC (2015) Anticipatory planning for courier, express and parcel services. In: Dethloff J, Haasis HD, Kopfer H, Kotzab H, Schönberger J (eds) Logistics management. Lecture notes in logistics. Springer, Cham, https://doi.org/10.1007/978-3-319-13177-1_25

Ulmer MW, Mattfeld DC, Hennig M, Goodson JC (2016) A rollout algorithm for vehicle routing with stochastic customer requests. In: Mattfeld DC, Spengler TS, Brinkmann J, Grunewald M (eds) Logistics management. Lecture notes in logistcs. Springer, Cham. https://doi.org/10.1007/978-3-319-20863-3_16

Ulmer MW, Heilig L, Voß S (2017a) On the value and challenge of real-time information in dynamic dispatching of service vehicles. Bus Inf Syst Eng 59:161–171. https://doi.org/10.1007/s12599-017-0468-2

Ulmer MW, Mattfeld DC, Köster F (2017b) Budgeting time for dynamic vehicle routing with stochastic customer requests. Transp Sci 52(1):20–37. https://doi.org/10.1287/trsc.2016.0719

United Nations (2017) World Population 2017. Technical report, United Nations, Department of Economic and Social Affairs, Population Division. https://esa.un.org/unpd/wpp/Publications/Files/WPP2017_Wallchart.pdf. Accessed 16 Mar 2018

Voccia SA, Campbell AM, Thomas BW (2017) The same-day delivery problem for online purchases. Transp Sci. https://doi.org/10.1287/trsc.2016.0732

Vogel P (2016) Service network design of bike sharing systems - analysis and optimization. Lecture notes in mobility. Springer, Cham. https://doi.org/10.1007/978-3-319-27735-6

Vogel P, Greiser T, Mattfeld DC (2011) Understanding bike-sharing systems using data mining: exploring activity patterns. Proc-Soc Behav Sci 20:514–523. https://doi.org/10.1016/j.sbspro.2011.08.058

Vogel P, Neumann Saavedra BA, Mattfeld DC (2014) A hybrid metaheuristic to solve the resource allocation problem in bike sharing systems. Hybrid metaheuristics, vol 8457. Lecture notes in computer science. Springer, Cham, pp 16–29. https://doi.org/10.1007/978-3-319-07644-7_2

Vogel P, Ehmke JF, Mattfeld DC (2017) Cost-efficient allocation of bikes to stations in bike sharing systems. Computational logistics, vol 10,572. Lecture notes in computer science. Springer, Cham, pp 498–512. https://doi.org/10.1007/978-3-319-68496-3_33

Wang M, Zhou X (2017) Bike-sharing systems and congestion: Evidence from US cities. J Transp Geogr 65:147–154. https://doi.org/10.1016/j.jtrangeo.2017.10.022

Wang Y, Szeto WY (2018) Static green repositioning in bike sharing systems with broken bikes. Transp Res Part D: Transp Environ 65:438–457. https://doi.org/10.1016/j.trd.2018.09.016

Wikipedia (2018) Mile. https://en.wikipedia.org/wiki/Mile (version of 2018-06-16, 6.21 p.m.)

Wooldridge JM (2015) Introductory econometrics: a modern approach. Nelson Education

Yan S, Lin JR, Chen YC, Xie FR (2017) Rental bike location and allocation under stochastic demands. Comput Ind Eng 107:1–11. https://doi.org/10.1016/j.cie.2017.02.018

Zhang D, Yu C, Desai J, Lau HYK, Srivathsan S (2017) A time-space network flow approach to dynamic repositioning in bicycle sharing systems. Transp Res Part B: Methodol 103:188–207. https://doi.org/10.1016/j.trb.2016.12.006

Printed in the United States
By Bookmasters